Chasing Smoke

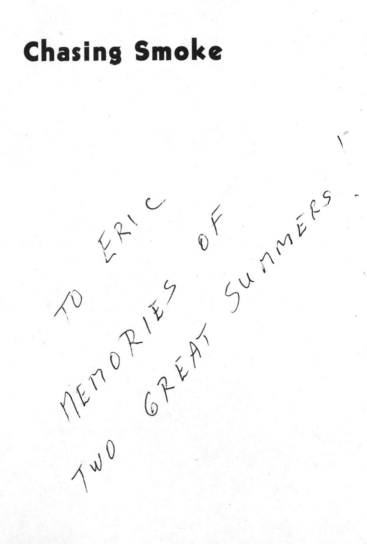

TO ERIC

MEMORIES OF

TWO GREAT SUMMERS !

Chasing Smoke

A WILDFIRE MEMOIR

Aaron Williams

HARBOUR PUBLISHING

Harbour Publishing Co. Ltd.
P.O. Box 219, Madeira Park, BC, V0N 2H0
www.harbourpublishing.com

Edited by Brianna Cerkiewicz
Text design by Mary White
All photos by Aaron Williams unless otherwise credited
Printed and bound in Canada

Canada Council   Conseil des arts   BRITISH COLUMBIA
for the Arts   du Canada   ARTS COUNCIL   Canadä
                                         An agency of the Province of British Columbia

Harbour Publishing acknowledges the support of the Canada
Council for the Arts, which last year invested $153 million to bring
the arts to Canadians throughout the country. We also gratefully
acknowledge financial support from the Government of Canada and
from the Province of British Columbia through the BC Arts Council
and the Book Publishing Tax Credit.

Library and Archives Canada Cataloguing in Publication

Williams, Aaron (Aaron Lloyd), author
    Chasing smoke : a wildfire memoir / Aaron Williams.
Issued in print and electronic formats.
ISBN 978-1-55017-805-0 (softcover).—ISBN
978-1-55017-806-7 (HTML)
    1. Williams, Aaron (Aaron Lloyd). 2. Wildfire fighters—British
Columbia—Biography. I. Title.
SD421.25.W54A3 2017   363.37'9092   C2017-905182-2
                                              C2017-905183-0

For Sue,
and for the Telkwa Rangers

Contents

Rangers' 2014 Fire Season

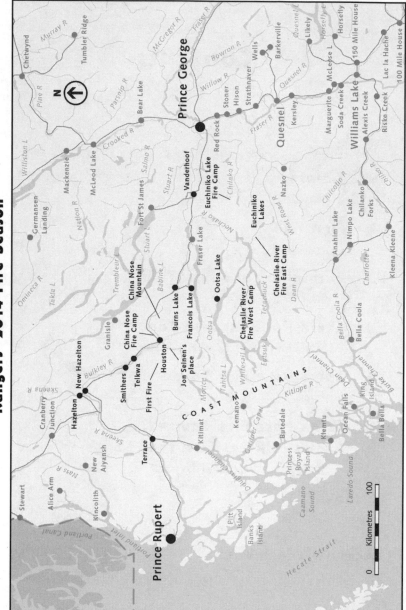

Prologue

The edge of this fire is supposed to be somewhere around here, but a fire this big doesn't have edges, at least not from the perspective of two guys walking down a dirt road in the dead heat of the afternoon.

Balsam and spruce trees candle on the hills around us. It's hard to distinguish the sound of distant burning from the gusts of wind shooting across this barren logging block. Brad and I continue along the logging road. We're looking for a small escape fire that has popped up on the north side of the road, the wrong side. This breach is one of many in the tenuous containment of the Chelaslie River fire, a massive blaze in north central British Columbia monitored by a few dozen firefighters and a few helicopters.

We see another road running parallel to ours higher on the hill. Above that, trees burn. They flare up in groups of two or three, the taller balsams being the most impressive to watch. Sheets of flame unfurl from their branches, sending black smoke into the sky to join the mothership cloud of grey smoke hanging permanently in the air above us.

We stare at the plumes of smoke and continue walking, frequently changing direction so it becomes more like pacing. It's impossible to tell what counts as contained here, and if some bit of fire isn't where it should be, it'll take more than two of us to put it out. With this in mind, as well as other factors including time of day (late), day of deployment (last) and general morale (low), we decide to walk back to our truck.

But as we're walking, a helicopter comes out of the haze, breaking the silence. It's Dan. He radioes us from the air, saying he sees a road for us to use. But we've already checked it and know it's a dead end. He says to wait there.

Ten minutes later, Dan arrives in the truck with two other crew members, Lauren and Kelly. He blows past the junction where Brad and I sit waiting. Seeing us as he drives by, Dan locks up the brakes, skids on the loose gravel and puts the truck in reverse. I can see from the slope of his shoulders, the jut of his neck, that he's enjoying the drama of his entrance. The chase is back on.

We find our escape fire on the next road up from where Brad and I had been walking. The spot is in open slash and it's active, churning through whatever litters the forest floor. The fire anchors itself to decaying stumps and root systems or flares up in the richer deposits of brush left behind by logging. The flames are over head height but there are places where they're less active, and from those areas we dig away at the edge of the fire, pulling it in on itself as if dabbing the edge of a wound.

A helicopter buckets another spot nearby, coming in and out of focus and earshot, disappearing into the smoke to refill with water at a nearby lake. The group of us, five in total, works in silence on different sections of the escape. There's laboured breathing and the clink of tools hitting rock.

Our bodies and clothes are filthy, our hands blackened and callused. The smoke cloud is starting to descend toward the ground. It's September 16, six days before official fall, way past its actual start in northern BC. Our season should be over.

Still, here we are, trying to contain the biggest fire the province has seen in thirty years.

1
Training

May 2014

Outside in the spring sunshine, everything perches on the edge of blooming. I'm heading to the Telkwa Fire Attack Base—"the base"—early to do some prep for our first week. I pull into a mostly empty parking lot. Around it is a flat forest of pine trees, beyond are the peaks of the Telkwa Range, a subsidiary of the Coast Mountains. Low on the hills are old clear-cuts, now covered with new trees. Higher up the mountains, fresh scrapes from recent logging hang above the original clear-cuts. The land is still brown and it looks like humanity changing its mind.

The first sign of life I find is inside the warehouse. I open the metal door and smell dirty clothes and gasoline. Rob and Warren are poking around in their lockers, getting ready for the day. The three of us are squad bosses on the Telkwa Rangers. Rob and Warren have the calm demeanour of tall guys with nothing to prove. They're veterans of the fire crew. I give each of them a hug, feeling the scruff of their faces close to mine. The three of us head across the yard to the unit crew trailer to plan out the day.

From inside, we hear the rest of the crew arrive. The river rock of the parking lot spits into the under-side of their vehicles, cracking and popping. Cars skid into parking spots, doors slam and shouted greetings cut through the surrounding pine trees.

We get our first look at the crew as they charge into the unit crew trailer. This trailer is the true home of the crew, made so partly by the people who breathe life into it, but also by memorabilia of years past. It's mostly old crew photos, but there are other odd heirlooms—a CD of the soundtrack to *Top Gun*, a stick from a deployment to Quebec. The first few to enter are quiet but as the group gathers at the two long tables in the middle of the room, each new arrival is met with more fanfare—applause, shouting, booing, anything.

A wiry guy with short blond hair comes in.

"Wil—ly," he says, drawing out my nickname, short for Williams, as if scolding me. This is how Brad, a third-year, has always greeted me.

There are only fourteen of us here right now, eleven guys and three girls. The youngest is twenty, the oldest thirty-two. Some are here for their second season; one, Warren, is here for his tenth. Missing today are five yet-to-be-hired rookies as well as our crew supervisor, Dan, who is at a training course this week.

The crew sprawls all over the generic office furniture. They're dressed in warm work clothes that are ripped and stained. Some of them bear the marks of coming straight from bed to the base—unwashed faces and gunk in their eyes. There are more hugs, but after warm greetings the conversation is oddly limited. I expect more, but intimacy will build over the summer.

We drive to Burns Lake for our start-of-year fit test later that morning. Two crew members, Tom and Mike, ride in my truck. We're two hours into the season and they're already imbibing in two commodities revered by firefighters—chewing tobacco and sleep.

Mike played a few games in the NHL as a goalie, and he's built like the hockey nets he once protected. He's reluctant to talk about his time in "the show." He's quiet and polite, and right now he's asleep in the back, his mouth slightly open.

Tom is built similar to Mike, but instead of a past in professional sports, he's spent a lot of his twenties partying hard. His physique depends largely on how much hiking he's done with his snowboard and how much beer and french fries he's been consuming. He stares out the window.

"My plan is to chew less this year," says Tom.

"Chew" is short for chewing tobacco; you take your chew in "dips." Tom is two dips in and we've only been at work for two hours.

"You're not off to a great start," I say.

"Yeah, I know man!" Tom says, laughing.

Up front I squint into the sun. I have a headache and I think it's seasonal shock—mountains, sunshine, the prospect of summer.

BURNS LAKE IS a mill town where the temperature is capable of dropping below freezing every month of the year. It's also a Christian stronghold. We've rented the rec hall at one of the churches so we can do our fit test. The church is just off the highway. Everything is painted sexually repressive brown.

In the parking lot is an assortment of high school kids hoping to set themselves up with a job after graduation. Once school is over, they'll be added to crews as part of a youth employment program run by the forest service. There are other prospective firefighters here too—area youths who might be granted a last-minute spot on a crew. Every year about fifteen hundred people apply to work as forest firefighters in BC. Of them, about two hundred are sent to an eight-day boot camp. Almost all two hundred are hired. But occasionally crews will still end up short a worker.

Outside the door, in the aftermath of her fit test, a pudgy teenage girl is wheezing. Around her is a circle of concerned onlookers. I hear the girl throw up just as I walk past. She does it all over herself and starts apologizing immediately.

None of us are on the verge of puking, but there are some nerves on the crew as well.

Before we're allowed to take the fitness test, we must take a blood pressure test. The test will keep our employer, the Ministry of Forests, Lands and Natural Resource Operations ("the Ministry" as we call it), safe from blame if somebody's heart stops during the fit test. But a blood pressure reading before a fitness test determining whether or not you have a job makes for a high blood pressure reading.

A third-year crew member, Chris, fails the blood pressure test. Minutes later he's lying down in the hallway trying to relax; he's pulled a shirt over his eyes and is listening to ambient electronic music on his phone. Chris is the fittest guy on our crew. He also contains more nervous energy than a first-time drug mule.

Later, he tells me why he thinks he failed: "I remember hearing somewhere that your blood pressure spikes when you eat, but for some reason I forgot and I ate so much food on the way to Burns Lake—I was just massacring food the whole way there."

Chris is a quiet guy. He's started balding but insists on growing his wavy dark hair, which still looks great in a hat. He has a dry sense of humour and lives most of his life in subtle, ironic detachment from whatever he's doing.

I get through the blood pressure test. After his relaxation therapy, Chris does too.

"The second time I took the test, I did it lying down," he says.

"Is that supposed to help?"

"I don't know. It worked, though."

The fit test is supposed to be a simulation of our actual work. Carrying a heavy backpack, we walk back and forth across the gymnasium, climbing up and down a steep ramp placed in the middle. Unlike the remote forests we work in, the ramp has metal railings for support.

The layout and the parts of the test are like a rodeo for humans. We don't get roped or anything, but we do have to pull a sled loaded with weights, which I would classify as a horse-like activity.

Our old fit test—a five-kilometre walk on a track followed by a series of sprints done while weighed down with hose—was used until 2012. We'd usually pass it with ease, but it was still competitive; there were bragging rights at stake. Many firefighters can recite all their fit test times starting from their rookie year. This new test is designed to stifle competition while still measuring fitness. They've done this by making it illegal to run. You can only do as well as your walking pace. Despite there being nothing at stake, crew members still walk as fast as possible, seeing how far ahead of the allotted time of fourteen minutes and thirty seconds they can be.

I promise myself I won't be competitive this year. I show I'm serious about it by leaving my wool shirt on for the test. But once I start, I can't help but reach further and further ahead of the clock as the person timing shouts out my progress: "You've done thirty laps. Twenty to go and you're forty-five seconds ahead!" The test ends with the heavy-horse pull event. I finish tugging and put down the rope. I'm the last to do the test, and we slouch en masse out of the gym once I'm finished.

Before heading home we get screamers at the Burns Lake Husky station. Screamers are a combination of slushy and soft ice cream. They're a catch-all remedy more effective than anything found in the drugstore. They quench hunger and thirst and help with things like joint pain and headaches. On the drive home I take huge gulps. Sunshine floods the truck and I smell the tin of chew, emptier than intended, being opened in the back seat.

IT'S OUR FIRST crew field day. Field days are a chance to test our gear and our ability to walk through the bush after a winter off. The plan is to use chainsaws to clear a short trail we've marked through the bush and then lay hose along the trail. Rob and I come in to work early to flag out a trail near our base.

I'll also be teaching everyone about water pumps. The

pumps we use are made by a company called Wajax. The model is the Mark 3. The stubbornness of these pumps is legendary. Everyone has a story about a Mark 3. I've seen them tumble hundreds of metres down a hillside and still run. They're also known to be cranky starters—their pull cords are stubborn, and bloody hands are a common side effect of troubleshooting a Mark 3.

The problem with my pump tutorial is I don't know much about pumps. It turns into a philosophy-of-pumps seminar: how to approach the pump, how you should treat it. I don't say, "This is how the piston fires and this is the fuel line." I'm not an authority on small-engine mechanics. Instead I say, "These pumps are very difficult. Be ruthless with them. If the pull cord bites flesh out of your hand, be even more aggressive." I don't know if it helps.

I keep thinking about Dan during the day. He's driving back from his course in Kamloops, away from the warm southern Interior and up into the still-cool north. I wonder what time he'll get home, and whether he'll work tomorrow or take the day off to be with his fiancée and their two young kids. I especially think of him while we're going through the pump exercise. Dan would have been all mechanics. He would have talked about the fuel, spark and air triangle.

When I picture Dan driving in the truck, I see his aura bursting through the windows. He drives fast and chews ice and listens to loud music. If he's on a stretch of highway with cell service, he makes phone calls.

One of them is to me.

"WILLY, YOU RAT." His voice blazes into my ear. Dan uses *rat* as a term of endearment.

"Hey man, how are ya?" I say with equal force.

"Are you guys ready to go?"

"Yeah, we're good." There's no need for small talk. I can't recall if we've ever small-talked.

The rest of the conversation consists of shouting and laughter. I smile at the phone after I hang up, as if it was what had entertained me.

I pull into the base the next morning and Dan is there, looking the same as always—running shoes, tattered shorts, sweater like it got caught in a lawnmower, greasy hair sprouting from the top of a ripped shirt sleeve worn as a headband. We greet each other with a hug and a laugh. It feels good, but I'm no longer the head of the crew, and I feel a pinch of loss as I follow him up the steps and into the crew trailer.

THE FITNESS TEST is over, but there's still plenty of fitness to dread. This is especially true early in the season, when we're eager to prove ourselves but not yet in very good shape. We're gathered at the Telkwa Elementary School, a five-minute drive from the base, near the head of a trail that winds through the woods along Tyhee Lake.

Ingrid runs our fitness program. She's a third-year who grew up in the Lower Mainland and was once on the rowing team at the University of Victoria.

She explains how we're supposed to treat the run we're going on. "It's not slow, but it's not a race."

"So it's an all-out race," I say, joking but not really. Fitness is always a competition on the Rangers. Even if we're supposed to take it easy, it becomes a contest to see who's taking it the right amount of easy.

"Let's go!" Ingrid says, and Dan barrels down the first hill at a full sprint, also joking but not really.

I start slower and fall in line as the not-race progresses, passing a few people, being passed once or twice. The trail is wet this morning, there are huge puddles where last year's leaves marinate in a cold stew. The crew charges through them like herd animals. Nothing but heavy breathing and the pounding of feet. The first one or two puddles are shallow. Then I hit one with some depth and the water pours into my shoes, browning my socks and turning my runners into wet foam strapped to my feet. A gap stretches out between those in front of me and those behind. I try to keep sight of somebody ahead so I have something to run for, but I can't help but let up a little when it feels like it's just me out there.

I come to the final stretch and hear shouts of encouragement from those who have already finished. It's these little things that make the crew what it is. I'm still a ways from any other runner, which is a relief because there will be no dogfight sprint to the end, which happens to somebody on just about every run. I finish fifth out of fifteen in the first (it's not a) race of the season.

After the run the whole crew has a meeting. The start-of-year talk includes writing out manifestos and creeds and goals on the whiteboard.

Dan does this meeting every year. The first one was when he took over as crew supervisor in 2007, transferring in from the unit crew in the nearby town of Houston. Many of his peers, people he'd grown up with, worked on the Rangers at the time. They found him arrogant and didn't want him joining the crew, especially not as their boss. After being introduced by upper management, Dan took over and started talking to his new crew. The mood in the trailer was sour and charged with defiance.

In the middle of his speech, Dan acknowledged the palpable hatred.

"I know you guys might not be so happy about some kid from down the street coming in and taking this job." He then added, "And let me get this straight, the streets is where I'm from."

I couldn't believe he did that—made a joke right when it looked like he was showing remorse to his seething peers. He had raised a white flag and then lit it on fire.

Dan left the topic of his unwelcome presence there, and that's where it stayed. He spent the next several summers rebuilding the crew in his image. Eventually, it worked. Having been on the Rangers before Dan arrived, I initially wasn't thrilled about the makeover. I'd spent my first year on a crew that was much less disciplined than the kind of operation Dan wanted to run. But I was young enough and Dan was charismatic enough for me to join his cause.

These meetings have long since taken on a different tone. Some of us, generally the more senior members, discuss what we can improve on from the previous season. The people to watch, though, are the second-years. We have four of them on the Rangers and they look relieved to be sitting in on this meeting. For them, last year was stressful: arriving at the base for the first time, an unknown job, a test of competency spanning several months.

FALLING IS THE most glamorous job on the unit crew. Crew members often don't get a chance to run chainsaw until they've been around for a couple of seasons. If you don't show some natural ability when given that chance, you'll be left behind treading water with the rookies. Being designated a faller takes even longer. There are six of us on the crew who are able to fall trees and this is our first day running saw this year.

After fifteen minutes, I'm comfortable. At sixteen minutes I'm fantastic. After seventeen minutes sweat pours off my brow, and even as sawdust clogs my pores I feel like I've been cleansed and reborn into a world that makes sense. The trees are running with sap this time of year and it coats my forearms. The sap mixes with sawdust and dirt, creating a series of brown splotches on my arms.

When I get home that night, I'm greeted by my roommate, Andrew. He and his wife, Sue, live in a small house in Smithers, a twenty-minute drive from the base in Telkwa. The three of us once worked together on the Rangers. They've since moved on: Sue now works for the Ministry on the management side and Andrew is an engineer. The night before, Andrew and I had talked about the good and bad of firefighting. The simple pleasure of getting through the day is something he doesn't have now; his work often comes home with him. His responsibilities are set to increase when he and Sue have their first child in September. Andrew catches sight of my dirty arms when I come in the door.

"I see you've been working," he says.

ON PAPER, LITTLE is demanded of BC's fire crews. We're protected by unions and anonymized in a stagnant bureaucratic pool where any small ripple would feel like a typhoon. The desire to do more—to find ways to help the community when not fighting fires and hold ourselves to a high standard—often comes from within the crews.

Crew leaders like Dan are trying to make us a more respectable organization and there are others in higher management who share this view. But in a job so protected, there are bound to be people whose greatest effort at work is dedicated to avoiding work.

With that in mind, today we're cutting a "fuel free," a ten-foot-wide break in the timber sometimes used in fire-fighting. Once a fuel free is cut and the wood has been cleared off (we call this "swamping"), we'll dig "hand guard" down the middle. Hand guard is a trench about thirty centimetres wide and five centimetres deep. With the right bird's-eye view, a completed fuel free and hand guard should look like a highway through a forest—a swath of cleared trees with a path right down the centre. In theory, a fuel free and hand guard should be enough of a break in fuel to slow or stop a large fire.

The whole process, from falling trees, to bucking them up, to clearing all the debris, to digging a hand guard, is a pretty insane amount of labour. And we're leaving the worst of that labour—the swamping and guard digging—for the rookies to do during their training week.

What's more ridiculous than the labour, though, is that we're cutting down perfectly healthy trees to teach our rookies the value of work for work's sake. I'll tell my grandkids this—about how we used to cut down living trees just to make somebody stack them in a pile. But for now, this is what makes us better. And any philosophical wondering about killing trees evaporates in haste as soon as I start my chainsaw. A primitive desire to harness nature rises up and tells me to mow it all down.

THE FIRST WEEK of work is done. It's an adjustment, one I've made many times and one that's getting tougher. The last two seasons I've left my winter home of Halifax and my girlfriend, Sue, who is originally from Smithers, to come back to firefighting. (Smithers is the kind of town where Sue remained a popular name well beyond its 1940s peak everywhere else.) I met Sue between my second and third seasons, and since then we've spent most summers apart—it wasn't so bad for a season or two, but it gets old. And with Sue now permanently working in Halifax and visits limited to as few as one per summer, it's getting old fast.

When I think back to previous first weeks of work, I can only recall being cold and feeling sleepy. Today, I lie alone in my room and stare at the ceiling. My eyes start to close.

"Willy, let's go." I open my eyes.

It's Andrew.

"Where?"

"We gotta do something, time is running out."

With the baby arriving in a few months and his Sue out of town for a training course, he's anxious to do some last-minute visiting with friends. I pry myself from my warm bed and we head out to socialize.

In the winter, one social interaction per week is about all my peers and I can fit into our lives. These interactions are usually quick and high energy. Here, the visits are long. Talk is slow and silences hang. Basic middle-class life is discussed—who bought which house, got what job, has how many kids now.

In this first week I've done more standing around the tailgates of trucks and tractors and equipment than I did all winter. Spitting and speculating on what's the best way to deal with a rust problem. It's like we huddle around this equipment because it's the only thing giving us a leg up on the vast nothingness that goes on forever in every direction.

ROOKIE WEEK IS a curated gauntlet of abuse meant to show new Rangers the worst of what they'll do as a forest firefighter.

Some of the abuse is paperwork. Most of it is meaningless, but it's an accurate representation of what they'll face on the job; for every five days of work, one whole day will be dedicated to feeding the bureaucracy.

Some of the abuse is physical. Today, the second day of rookie week, we've headed to the field after our morning workout. The rookies are swamping and digging hand guard on the fuel free we cut last week.

While the rookies work, I do some chainsaw instruction nearby with one of our second-years, Lauren, who's getting her first bit of saw training this year. Lauren was on my squad last year. I found her difficult; she was a bit older than the average rookie and had strong opinions.

Today she's fine, though; the saw is a good challenge for her and she mostly works on her own.

I feel chippy today. Maybe it's residual anger at not getting a promotion I applied for this year. It was a crew supervisor position, what Dan does, on another crew. I was passed over for somebody I thought wasn't as qualified. But maybe my sour outlook is just due to all the years I've been at this. It's my ninth season. I have two arts degrees and no hard skills. I'm still on the Telkwa Rangers unit crew. Seasonal employment. The government teat.

At the end of the day, the rookies are still digging hand guard. They're slow and they don't know how to swing their axes. This group that was so keen earlier in the day is lost in the milk-and-cereal simplicity of digging guard.

Meanwhile the rest of the crew looks on, yelling words of encouragement and talking excitedly among themselves. Our goal with new people is to be nice to them but also to make them work hard during rookie week. I put on as much of a surly manner as my congenial Type B self will allow. I don't make eye contact; I try not to laugh. It's an affectation, and probably useless. Still, it's good to set low friendliness standards in these situations.

Anger takes hold as I watch the rookies dig guard. I tell them to pick it up. They look like sad cafeteria employees heaving food onto plates. The crew continues to chat, the din getting louder, the laughs harder. It creates a thunderstorm atmosphere. Some of the veterans look for more effort while others are happy just to be out there relaxing and cheering on the new guys. I don't want to work beside some new guy who has only been applauded when we're on a real fire and under the gun. I want to be next to somebody who pushed it and felt fear during their rookie week.

WE ARRIVE AT a trailhead on the side of a logging road that follows the Telkwa River. The trailhead is a wide, water-filled ditch. Getting wet can be avoided if you're careful and take a detour. But in the spirit of keeping all rookies miserable, we hint that the best and fastest way up is to just walk through the ditch. From there the path climbs for about five hundred metres until it hits the flatter ground butting up against the base of the surrounding mountains. Parts of the trail are so steep that vegetation has a hard time clinging to the soil. On these steep stretches the trail turns into a minefield of ruts, some of which are too soft to support the human body plus almost a human body's weight in gear.

We unload the fire gear from the back of the work trucks—pumps, gas cans, chainsaws, hose—and stack it at the side of the road.

Dan gives a short speech to the rookies. "Okay, gang, this stuff has to make it to the top of the hill. Trail is right there."

The exercise comes off as unnecessary and cruel, but this is one of the most applicable parts of the week. When we walk into the bush to fight a fire, we're almost always carrying something heavy. It's normal to head into the bush with more than forty-five kilograms of gear.

Everyone doing the gear carry will someday have a harder day in their life, but when you're dealing with nineteen- and twenty-year-olds, the standards for a hard day can be missing a Fruit Roll-Up from your lunch kit.

The rookies react to the gear carry with the stunned look of an animal with an arrow in its side. Some withdraw into themselves or look angry at us for hurting them. Others change their expression, their faces twisting in a way that says they understand this is life. Sometimes you have to haul two twenty-kilogram gas cans up a hill while your co-workers watch.

The rest of the crew spreads out along the trail to watch and encourage the new Rangers. I find a spot near Dan and Chris. Dan talks about the gear carry in his rookie year in 2001 on the Burns Lake unit crew.

He laughs about some of the things they had to haul up the hill—a tool box full of rocks and a half-eaten pizza. They used a similarly steep trail, but in order to avoid the worst of the climb some of Dan's fellow rookies decided to stockpile the gear on the trail just before it got steep. Dan and a few others were stuck taking the gear from the base of the hill to the top.

"We ended up doing that steep pitch over and over again while these other folks"—he pauses for a second, trying to find a way to be diplomatic—"missed the opportunity to earn our respect. That's part of the whole equation. Each one of these people needs to pull their weight all the way to the top and earn the respect of their peers."

He's right. But as is often the case with Dan, his sermon and his views on firefighting are nearly crushed by a seriousness that even jokes about carrying pizzas up steep hills can't negate. Chris stands and listens; his face is a blank slate. I'd love to know his thoughts.

Before I continue finding things to nitpick about Dan, I should stop and say that although our relationship can be strained, it has good roots.

The first time I met Dan was at firefighting boot camp in 2006. I was a rookie and he was an instructor. I spent as much time around him as possible. He represented the exception to my early interpretation of the forest firefighter ethos—egotistic, "extreme" dudes who wore sunglasses. Dan was so far the other way that he wouldn't let anyone on his crew wear sunglasses.

Dan made such an impression on me that weeks after camp ended I was still actively trying to be like him—talk like him, act like him, even carry myself like him, which meant with confidence. Too much confidence, but something I needed at that point in my life.

The rookies walk by, the initial variety of facial expressions shown at the beginning having transitioned into a homogeneous look of vague pain. We continue to encourage but we're talking to them while they're underwater.

Farther up the hill, I run into Addison, one of Dan's more recent recruits. Up here, the hill turns into a razor-back ridge dropping off sixty metres on each side. The wind on this ridge blows directly off the glaciers in the mountains. The air is so fresh it reaches your toes when you breathe deep.

Addison is a second-year. We didn't talk much in his rookie year. I saw him as overconfident, always putting on a show for the guys. This year has been better between us. This is how it is in firefighting—sometimes people get more bearable simply by sticking around another year.

Addison's voice is low and raspy but also sweet-sounding. He's the youngest of four boys, a tough farm kid from the north. But he also has a strong sense of empathy and a shrewd understanding of how people socialize and how they judge each other.

His family has two thousand acres of land and five hundred head of cattle, and they live about ten kilometres from our base. Their property sits on a road named after their family. By regional standards they've been there forever.

Addison's father is getting old and Addison stays on the farm in the summer to help with the work.

"The old man is crippled," he says. "He went to the doctor and the doctor said the only reason he's still able to stand is that he's completely worn out the nerve endings in his knees."

"He used to wake up at three a.m. to go logging, and then come home and work the farm," he adds.

Addison gets up early to help his parents, works a full day with the Rangers and goes home in the evening to tend cattle.

A rookie named Kelly passes us on the hill. Her expression is somehow vacant and miserable at the same time.

"Good work, Kelly," Addison says, almost in a whisper.

Kelly later tells me her thoughts on the gear carry. "All day I couldn't stop thinking, 'It's not worth it, no job is worth what I'm doing. I could be in a café somewhere eating doughnuts.'"

Addison waits until Kelly is a ways up the hill before we continue talking. He says some people don't understand work and that this rookie week is their first real taste of it.

Addison talks about the Ministry's preference for hiring high-performance athletes. In the winter, interviews are held at a number of BC universities, where members of school teams are given high priority.

"Good athletes and good workers don't have much in common," he says. "Hard workouts twice a day isn't the same as going to work for twelve hours straight. It takes a different strength to work those long days."

Addison used to work concrete in Edmonton. That's what he calls it: "working concrete." He'd work fourteen- and fifteen-hour days bent over fresh concrete, running trowels over long sections of sidewalk. Addison believes it's the unassuming guys who are the best employees. He thinks about these things—about the nature of manual labour.

After work, Addison is in his farm gear. He has a denim shirt on, faded jeans with green-brown earth worked into the thighs, suspenders and a pair of old leather work boots. He's a ranching sexpot.

WE DRIVE NORTH into the Babine Mountains and turn off a wide dirt road onto a narrow one littered with debris, clippings from previous windstorms. Today is water pumping exercises. A few kilometres up the road we park the trucks within earshot of a large creek that's wailing under the stress of the spring melt. It's cooler up here than in the valley bottom and coolest near the creek.

Yesterday the rookies were obliterated by the gear carry, and now things are getting ugly. They walk like they're barefoot on broken glass. They wince if they have to bend over. Even their heads seem to bow under the weight of their hard hats. Today's work involves strategy; it will be more mind than muscle. But the muscles are so worn out that they're affecting the minds. The rookies are terse with each other as they try to figure out their task, which is to find a water source, set up a pump and string hose out to an area we've designated as "the fire." They'll go through the drill a few times, the idea being to get faster each time.

The day is long and a group of vets stays out with the rookies until it's nearly dark. We stand in a circle and chat, each taking a stab at entertaining the group, seeing who has the best story or who can deflect insults the fastest.

Dan talks about all his commitments and debts—a

rental house in Telkwa, another house in Smithers, a fiancée and two kids under the age of four. He wishes he was nineteen again.

"I had a truck and I could put all my things in my truck, and I just wanted to have sex," he says. "Sometimes you wish you could go back to that simple living."

After a second I say, "Yeah, well, it's not that sweet when you're twenty-eight and you're still doing it."

Big laughs from the group and I laugh as well, unable to contain my pleasure in causing this reaction.

By the time the rookies finish, it's late. Dan and I drive back together. It's ten at night when I finally get home.

SEVEN HOURS LATER, at five a.m., Dan is parked outside my house again. I struggle out of bed, put on yesterday's clothes and grab the remainder of a carton of chocolate milk and a power bar from the cupboard. We talk quietly in the dark interior of his old truck. The leather upholstery is popped and bleeding foam everywhere from hauling around dogs and tools.

"How was your night?" Dan says.

"Good. How was yours?"

"Not good, man. Moses got run over."

Moses is Dan's dog.

Dan came home to find his fiancée, Jen, crying on the outside steps. They looked at each other and Jen said, "I'm sorry, I'm sorry." Dan knew what had happened.

"Where is he?"

"He's in the shed."

He got a shovel and at 10:30 p.m. started digging a hole to bury his dog. He didn't finish until midnight.

At work, Dan tells no one else.

It's the last day of rookie week; the sky is pure blue and the temperature climbs as soon as the sun comes up. Today we'll be around real fire. Visible from our base is an abandoned airstrip a kilometre long and a hundred metres wide. On the south end of the airstrip a few crew members have shown up early to get some fires going. They're burning debris piles but allowing the fire to creep away along the forest floor. Residual dampness from the spring melt means there's no threat of the fire getting out of control.

From the north end of the airstrip we see sheets of blue smoke coming out of the trees. I'm in a truck with a couple of rookies. They're tired but their faces light up with surprise and worry as we drive closer to the burn. I feel it too. Smoke—even contained, docile mid-May smoke—makes my heart race.

The first time I had this feeling was long before I started firefighting. I was about eight or nine and was on holiday at my grandmother's cabin on a lake. The cabin had a firepit, and a bonfire from the previous night smouldered in some bigger logs. In the middle of the afternoon, while I was playing on the beach, black smoke started billowing from the firepit. The fire quickly got into a nearby clump of bushes and for one terrifying minute I pictured everything going up in flames. Not just everything in my immediate surroundings, but every-thing in the entire region. It was naive but also, I believe, one of the things that scares people about forest fires—where do they stop? Even big structural fires you see on the news, burning warehouses and apartment build-ings, tend to be hemmed in by adjacent parking lots and streets. Not forest fires—they have room.

So when we roll up to an arena-sized swath of forest roiling under a low-intensity surface fire with no natural barriers to run into, it's easy to be deceived into thinking we're seeing the beginning of a situation that will one day be the subject of a tragic Heritage Minute. But taking stock of a few things puts it into perspective: we're in northwestern BC, it's May, it hasn't been particularly hot lately and it's early in the morning. If I came upon this scene on a mid-July afternoon after two weeks without rain, there would be problems.

The rookies learn as much at boot camp—how weather and the type of forest and the time of day and year all affect a fire—but taking everything into account and coming up with the right plan is harder than reading about it. The rookies are slow, but they do all right. They set up a pump at a nearby pond, lay out some hose and start putting water on fire. I wouldn't yet trust them to protect my house. Maybe my shed.

When the practice fire is out, the rookies have one last challenge. They have to do the "red roof run," a trail frequented by local rednecks on ATVs. The trail runs through pine flats near a dead-looking swamp, which makes it muddy for long stretches. They'll do the run while together carrying a piece of inch-and-a-half hose full of water, weighing about seventy kilograms. They'll run with the hose over their shoulders. When they reach the muddy ruts near the swamp, they'll stampede through them, the mud sticking to their shoes, making them heavy and slippery. After the mud they'll hit a big mother of a hill. By the time they get to the top they'll be delirious. Dan loves this kind of sports movie finish: the group coming together for one final challenge before they're welcomed onto the crew.

During the run one of the rookies, Brian, starts to fall apart; before they hit the big hill his expression changes from tired to something worse. The sheen of sweat disappears and his eyes are alive with panic. His legs are covered in mud and he has festering grey athletic tape wrapped around one of his ankles. Dan catches his exhaustion and takes his load for small stretches. Just after the top of the big hill as the road starts twisting back to the valley bottom, Brian veers off to the side of the dirt road and collapses near the ditch.

The group keeps running, downhill and out of sight with the hose flapping madly along like a kite tail in the wind. Warren and I stay behind. Brian is on his knees, head down so low his lips almost touch the dirt. He convulses with each breath. The lack of sweat sets me on edge. Warren tells him not to worry, to take deep breaths.

"I'm so sorry," he says. His head still hovers just above the road.

"Don't be sorry," Warren and I say at the same time.

"I let down my team," he says.

He gets up and the three of us start walking. The height of his suffering has passed and now he needs to make amends. He covers the tracks of what went on just fifty metres behind us by making small talk. "Will we be on standby this weekend?" he says. We come around a corner and hear the whirring of the air filters at the Telkwa tree nursery. We're still up high enough to have a view of the valley. The trees, with their tufts of spring foliage, shimmer as they sway in the wind, making it look as if the whole valley is shuddering.

We catch up with the rookies carrying the hose and arrive at the base, where the rest of the crew is waiting in the driveway to cheer them on. The rookies shrug off the hose. Some squat on the grass, catching their breath. Others lie on their backs and close their eyes, their heartbeats visible through their shirts. The veterans offer hugs and handshakes.

This is the 2014 incarnation of the Telkwa Rangers, one of the first three unit crews put together by the provincial government in 1988. Ours is one of the only crews in the province whose founding mothers and fathers could also be our actual mothers and fathers.

Dan gives a short speech.

"In this job you may find yourself in muskeg, soaking wet, covered in bugs, camping on the line. We pride ourselves on being able to laugh at the shittiest of the shitty," he says. "The bottom line is, we're going to try and be our best selves…as individuals and as a crew."

ASK ANY FALLER what it was like the first time they cut down a tree and you'll get the same answer: it was scary.

The first time I fell a tree was in 2007 in a desolate cutblock in the mountainous southeastern corner of BC known as the Kootenays. We were working on a quiet fire in early autumn. The workload was easy and Dan had sent a few of the newer people out to do some falling practice.

I was partnered with Brendan, a squad boss who grew up in Smithers. He was generous with his laughter, which went a long way for a new guy like me trying to fit in. He was the first on the crew to call me Willy.

After giving me some pointers on falling, Brendan stood at an I-trust-you distance when I cut into my first-ever tree.

I took my time with what's known as the undercut—a

cut that involves opening up part of the tree's stem. If you picture a whole watermelon and how you might extract a slice without cutting through the melon, that's an undercut. After the undercut I worked on a back cut—a straight slice through the stem. Once the back cut and undercut are done there should still be a thin strip of wood connecting tree to stump. This strip creates a hinge that helps control the tree's direction as it falls.

I was slow throughout the process. Thankfully, my clumsy form and lack of awareness were forgiven, as I had picked the perfect tree for my first stab at falling. As I worked through my back cut I sensed the approach of the crucial moment when the tree begins to fall and the chance of gruesome death reaches its apex. The tension rose until I heard the crack of the tree shrugging off the holding wood and succumbing to gravity. I slowly walked backward, eyes on the falling tree, heart pounding in my chest.

Success with the first tree boosted my confidence. I started into the second one, going through the same motions. I reached the end of my cuts and stood back when the tree was supposed to fall. It didn't move. I stepped up to the tree again. Was my strip of holding wood too thick? Was it still anchoring the tree to the stump? I started my saw and shaved off more holding wood. At this point I should have used a wedge—a tapered piece of plastic that can be placed in the back cut and pounded in with an axe to assist with falling. Instead, I kept shaving off my holding wood, bit by bit. Why wouldn't this thing fall? Like a true novice, I cut through the last fibre of wood holding it together. For one queasy second the whole tree swivelled around on the stump—like a coin dropped on a table—threatening to fall in every direction. It then decided to (thank God) fall roughly in the area I had been hoping, keeping me clear of danger.

That second tree spooked me, showed how illiterate I was when it came to reading trees. It had been dead before it burned, which created a massive change in how it reacted to being felled—one of the many aspects of falling I didn't understand.

Today, Brendan will watch me fall trees again. This time it's for the Ministry falling exam, an annual checkup that I'm doing for the sixth time. Since that day in the Kootenays in 2007, Brendan has moved up from squad boss to "protection assistant." He does things like administer falling exams and decide how we're going to fight large fires. I'm now in my fourth year as a squad boss. The drive we're taking is a welcome break for both of us. For me it's time off from the tension of prepping an inexperienced unit crew to fight fire. For him it's a break from endless emails.

I'll do the falling test on a logging road southwest of Houston, a mill town of four thousand people a half hour's drive from our base in Telkwa. The road we're on is plied by logging trucks that serve one of the biggest sawmills in the world. The landscape is a patchwork of features, from healthy green rivers to muskeg wetlands. As it would in any direction, our southwest tack puts us deep into a remote wilderness. We pass Nadina Mountain, the most imposing feature in the area: a stand-alone peak rising 1,800 metres from the surrounding pine forests. On the plateau of Nadina Mountain, there's a snow-melt pond, ice-free for a couple of months in the summer. One year, Dan and Brendan were in a helicopter over top of the mountain, heading to a nearby fire. Floating in the middle of the pond was the carcass of a moose. Standing on top of the moose was a wolf tearing at its flesh. The rest of the pack looked on from the edge of the pond.

I tell this story because a) it's cool and b) it's not shocking to hear details of this kind of raw apex predator activity in this area. In most of the world, it's a treat to see an animal. Period. But around here, nature stories are only worth telling if you've witnessed animals engaged in graphic acts of survival.

About fifty kilometres down the road, we arrive at an area that was scorched by several massive fires in 2010. Nearly fifty thousand hectares were incinerated along this road system. The combined aftermath of logging and forest fires makes the Earth's surface look like food burnt onto the bottom of a cooking pot. I'll do my test in this fragile, blackened timber.

Before we start, I realize I've forgotten to bring gas for my saw. Brendan lends me his, but the gaffe makes a nervous situation worse. I've gotten better since that day in the Kootenays years ago, but fallers in the Ministry are never great. If it's a busy summer we'll spend a couple of months falling—"time at the stump," as it's known. But we're nowhere near the level of our counterparts in the world of production falling. Production fallers usually work in the logging industry and their job is solely to fall trees.

I start by taking a couple of practice trees down, badly misjudging the levelness of my cuts. I calm down, though, and with each successful tree my confidence builds. By the end Brendan is challenging me to cut down more difficult trees, an indication that I've passed.

When we finish, we eat borscht in his truck, hiding from the rain and single-digit temperatures.

It's late May and fire season is nowhere in sight.

WHEN I INTERVIEWED to be a firefighter in March of 2006, I didn't know anything about the job. I came from Prince Rupert, a rainforest so wet it could repel napalm. My lack of knowledge was obvious throughout the interview, most obvious when the guy conducting the interview, Dave, asked, "What do you think we do in our downtime?"

"I don't know," I said. "We probably do maintenance on our oxygen tanks."

I didn't know that oxygen tanks were in no way part of the forest firefighter's equipment. Using an oxygen tank when fighting a forest fire would be like using a football to play basketball.

So there were no oxygen tanks, but almost nine years later, Dave's question still lingers—what *do* we do in our downtime? We have lots of it. We usually only work on fires for about half the summer.

When we're not firefighting, we do "project work," which ranges from cutting firewood to brushing forest service roads to laying bricks. Sometimes it's meaningful, but it always feels a little forced at best, and like a pointless drain of tax dollars at worst.

The first of this year's project work is burning brush piles near the base. This kind of work is called "fuels management," though nobody adds the *s* to *fuel* except for the managers who oversee the program. Fuel management took shape after the 2003 fire season, when everyone and their military was trying to put out fires in the southern half of the province. It was the year the

Okanagan Mountain Park fire burned nearly 250 homes in a suburb of Kelowna.

Fuel management is an ongoing project. Every summer we cut down dead trees and limb low-hanging branches on green trees. Afterward, we pile them up in one spot and, when the conditions are right, burn the piles. This is to stop "fuel loading," that is, the accumulation of too much burnable stuff on the forest floor.

The size of the project is beyond comprehension and, as far as I know, fuel management hasn't been proven to actually stop fires. More than a decade into it we can still see our fire base from where we've been picking up sticks.

Partway through our morning of pile burning, a guy cruises up in an old pickup. It's a local checking to make sure our burning isn't getting out of control. This happens often. The service we provide the people mostly just sets them on edge.

At lunchtime I sit with Kara. There are bonfires burning all around us. We sit on decaying logs, bits of broken glass from old bush parties glinting in the sun.

Kara is a second-year who grew up in Smithers. She's in her midtwenties and has just finished university. Her mom is from Papua New Guinea and her dad is from Germany—an unusual combination in blindingly white Smithers. She's at ease in the testosterone-heavy environment of the unit crew.

There are usually three or four women on the Rangers, in step with the average. Of those people who fight forest fires in BC, roughly 20 percent are women.

My tone changes when I'm one-on-one with somebody, more so if it's with one of the women on the crew.

I can drop out of the endless subtle pissing match that is working with a group of guys.

Kara says she broke up with her boyfriend last week.

"That's it," she says, as if saying it out loud will convince her it's real.

"Man, things are going to get better," I say. "We're going to get on the road this summer, out working, making money, we'll..."

I want to tell her that meeting someone new if you're a woman doing this job isn't exactly difficult. But that seems reductive.

"We'll be having a great time with the crew," I say.

There's a heightened sexuality that prowls the edges of fire crews. Any activity this taxing and fitness-based is bound to be a lighter at the fuse of your sex drive. On the Rangers this is discouraged, and every year Dan throws a verbal bucket of ice water on the subject of "love on the line," as he likes to call it. Nonetheless, there's always a fair bit of flirting and just about every year there's some actual crew romance. Ingrid and Brad are together right now. The most Dan can hope for is that it's subtle. There are crews where this isn't the case and the result is a group with an adolescent atmosphere that gets under everyone's skin. Crew mates holding hands at work. Disgusting.

We finish up our lunch. The wind is strong in the afternoon, and the blue smoke is thick among the young pine trees. We're in the heat constantly, chucking green rounds of wood onto the piles, rounds weighed down with water and sap. Each pile hisses with the sound of water being squeezed out of wood. Near the end of the day I stop and lean on a tree. My arms are burnt, my nose is running and my eyes are raw. The tree I'm leaning on moves in the wind and I can feel the roots under my feet lifting me up and down a little, like I'm standing on a dock hit by a boat's wake. We're barely hanging on.

WE'RE CLEARING SOME hiking trails near Smithers. The crew is lethargic. A big group operates brush saws, but the blades often glance off rocks and they get so dull they emit a burning smell as they try to cut through the brush crowding over the trail. I'm raking the debris left by the brush saws; just ahead of me Tom is doing the same.

These are what Andrew calls the June Blues—the period in firefighting when we're bored of project work and antsy for real fire season to begin. We've been at it for nearly six weeks now. Fire season should be here.

June is also an important month in terms of determining how the bulk of the fire season will pan out.

A protection officer, Brad, explains this to me when I stop in at the Northwest Fire Centre one day after work to talk about fire weather. Protection officers have a variety of specialties; Brad's is weather and what it means for fires. We're talking in an office compound next to the Smithers airport that houses all of the senior employees in the Northwest Fire Centre.

Brad is in his midforties and wears hiking boots with blue jeans. He started his career almost thirty years ago in Dease Lake, a legendary hellhole a few hours south of the Yukon border where winter lasts for ten months of the year.

Brad loves forest fires and has a flair for talking about the weather that produces them; it's like listening to the origin myths of different cultures. The systems we contend with in BC are the "four corners ridge"—a high-pressure system that builds over the southwestern desert of the US—and cold low-pressure systems that form in the Gulf of Alaska and empty out over the Coast Mountains.

"Cold lows generally come in June," says Brad. "Anyone who's been around will tell you that June weather affects our fire season. Valley bottoms will dry out no matter what, but higher elevations take a long time to dry out."

Because the province is so mountainous, there is potential for forest fires at many elevations. High-elevation fires are more common on years with early snow melts and little rain.

"When we get our ass kicked, it's at all elevations," Brad says.

He explains that Smithers "sits behind the waterfall" of the Coast Mountains, which catch most of the precipitation from the Gulf of Alaska lows. From a window in the building, Brad points out where this figurative waterfall ends, which is about forty kilometres to the northwest. His talk is well timed because through the window I see the interior side of the Coast Mountains; they're covered with dark clouds and taking a solid hit of precipitation.

The following day, back at the hiking trails, we find ourselves under the waterfall Brad said wasn't our problem. It hails in the morning. It rains in the afternoon. I spend the day bent over, raking leaves with my fingers, throwing them into the bush behind me. Ahead, Tom scrapes the trail clear with a rake, moving at nursing-home pace.

The June Blues are worsened by how pointless being away from Sue feels when we're not even working a fire. We've been apart for less than two months and already

I'm feeling the long-distance misery, especially in the evenings. My only strategy is to try to get to bed early; the mornings are always more hopeful.

IN MY HEART of hearts I'm not a fight guy, I'm a flight guy. This is problematic because today we're doing our old fit test, and this test is a graceless fight to the death.

The old fit test is the biggest fitness event of the Rangers' fire season. Because it's not the government-sanctioned test anymore, we don't have to pass, but we don't have to pass in the same way other fitness we do "isn't a race." The winner gets their name and test time felt-penned onto the handle of a Pulaski, a combination axe/grubber we use at work. The handle is then drilled into the wall above that year's crew photo. In a job that doesn't have much room for legacy, winning the Pulaski handle is like having an international airport named in your honour.

There are two parts to the old fit test. The first is called the pack test—a twelve-lap (roughly five-kilometre) walk around a track with a twenty-kilogram backpack on. The second part is the pump-hose test, where we carry pieces of fire equipment back and forth in a straight line. These pieces are a pump, four lengths of rolled hose and one length of hose filled with water, which we drag along the ground. The distance of our suffering is four hundred metres. The depth of our suffering is far greater.

I first heard about the fit test when I was invited to boot camp in 2006. I trained for the test by running a five-kilometre trail in Prince Rupert wearing a backpack stuffed with two twenty-pound dumbbells wrapped in blankets.

I was in terrible shape. From a young age, organized sports had been a year-round part of my life. But when I finished high school, the practices and games ended with it. It was a shock running that trail with that big backpack pushing me around from behind. I didn't understand how my fitness could have gone downhill so quickly. But it had, and boot camp saved me from the flabby, direction-less state of my life.

At boot camp we were told we were only allowed to walk for the pack test; the pump-hose portion was where the running took place. I was apprehensive, since I had been training by running. But there were no hills on the track, so maybe it would even out.

Boot camp was held in the small town of Merritt in the southern interior of BC. We did the fitness test on a track near the local high school on a cool, cloudy morning in late April. Our test group had about twenty-five people in it. The pack test started without ceremony. One of the instructors beeped a stopwatch and yelled, "Go!" All of us newbies stopped talking and started down the straight stretch of the track, trying to figure out how fast we had to walk to stay ahead of the forty-five-minute time limit.

A few laps in, I'd slotted myself into third place. The two people in front of me were a tall, thick guy whose limbs were all roughly the same girth and length, and a skinnier guy who had shown up to boot camp in an old blue Camaro. He drove his Camaro to the track at the Merritt high school even though everyone else rode with the instructors in Ministry trucks.

I was proud of myself and my third-place spot. Little did I know the instructors didn't give a damn about how well I was doing on the pack test. The pack test is just used as a strength zapper to make the more prestigious pump-hose test extra brutal.

I was coming down the final stretch of my last lap, and was about to lap a big group of walkers, when the whole group, about a dozen of them, crossed the line and raised their arms in celebration. They had only done eleven of the twelve laps, but somebody miscounted and I went from third place to fifteenth. I didn't call their bluff, as it would have put me on the outs with everyone at camp, but it was hard to act like everything was cool when I crossed the finish line a minute later.

The pump-hose test came next. This was scarier. I hadn't practised it on my own.

I didn't go first and what I saw was disturbing. People started out fine but kept becoming more and more tired until, during the last portion (dragging the hose filled with water), they were bent right over so they were almost parallel with the ground. Some fell to their knees and crawled for stretches. People were coming across the line breathing as though their lungs would turn inside out. This was going to be the most difficult physical test I'd ever done. I'd done tough stuff for sports, but that was with friends and I was never at risk of being cut from the team. Now I had a job on the line.

The test had to be done in under four minutes and ten seconds. A few people failed before I went, most of them women whose body weight barely exceeded the weight they had to drag and carry. The most shocking fail was Camaro Guy, the rightful pack-test champion. He started strong but ended up totally gassed and falling all over himself in the second half of the test. After he failed I couldn't look at him; it was too sad. I'll never forget watching Camaro Guy fire up his Camaro and pull out of the Merritt high school parking lot. I wondered if it wasn't the first time this sort of thing had happened to him.

To my surprise, my first crack at the pump-hose test didn't end in disaster. By the time I got to the hose-dragging portion, I wasn't moving fast, but it also didn't feel like I was wrestling a great white shark. I finished in two minutes and forty-two seconds, an average time for an average boot camp attendee.

Today we're at Chandler Park Middle School for our test. The school closed fifteen years ago; the windows are boarded up and weeds are taking over the property. It sits just off the highway coming into Smithers from the east. There's always a parent or friend who drives by during the test and lays on the horn. Someone from the local paper usually makes an appearance too. This is an event.

Next to the school is a track and two soccer fields. The track has been covered with loose gravel, which is like building a swimming pool and filling it with syrup. We kick up dust when we start, and its smell spoils the fragrance of grass and morning dew that greeted us out of the trucks.

By lap two of the pack test, the sweat seeps out of my forehead; by the fourth lap, my shins are pulp. Walking fast in workboots on loose gravel is torture. With every step it feels like we lose half of our forward momentum.

Soon the twenty of us are spread out around the track. Some people talk and walk, most do it in silence. Everyone walks with their head down, eyes fixed on a spot about a few feet in front of them.

I feel particularly beat after this pack test, but it's probably nerves for the upcoming pump-hose test. I'm second in line for the test and I'm lined up against rookie Brian. Though Brian hasn't shown he's particularly fit, I'm still worried he's going to beat me.

The timed portion of the pump-hose test doesn't start right away; first you have to walk thirty metres with a thirty-kilogram Mark 3 pump. This tires us out and prolongs our dread of what we're about to do.

Brian and I finish our walk with the pump at the same time and I feel my throat tighten, robbing me of precious oxygen. We exchange the quickest of glances to make sure we're going to drop our pumps at the same time. The pumps hit the ground and the timer starts.

We're tied through one lap and I'm pulling harder than I usually do on the test. I don't want Brian to beat me. It's humiliating to get beat by a rookie in anything, more so on the pump hose.

Early in the test, while carrying the four lengths of hose, I realize I have about five steps on Brian. Go hard now and widen the gap. By the time I start dragging the water-filled hose, I'm ten steps ahead. I start with the hose and it's lighter than it's ever been. I pump my legs and there's still something left. This makes me feel incredibly strong and fit. I hear yelling but no words. It sounds like a crowd recognizing effort. I turn for my last lap and keep giving it everything. By the time I cross the finish line the yelling is more distant. The tone of the yelling, which is all you can process after the test, has changed. "Good effort but nowhere near a winning time," it says.

Two minutes, thirty-six seconds is what I get. I'm triumphant for a brief period. I'd always wanted to get into the 2:30s. After a minute of lying on the ground, eyes closed, chest heaving, I get up to cheer on the others.

Walking around, I can see my shadow on the grass; it's the shadow of a fit guy, an athlete. I continue to cheer on the crew, and many of them hit better times than me. Five of them do it faster than two minutes, twenty seconds.

The real showdown, though, is between Chris and Brad. It's too bad they're not actual rivals because in many ways they're perfect opposites.

Brad is a tough, working-class hockey player. Like Chris, he's about six feet tall and there's nothing on his body but muscle. Brad's parents split up when he was young. His older brother Wayne is one of Smithers' most famous sons. Wayne, four years older than Brad, was a top-level pro mountain biker for most of his twenties. Between the broken home and the famous brother, Brad is obsessed with proving himself, but he doesn't do it in a problematic way. He's gracious in defeat and victory. Chris is too, and so the rivalry lacks substance. There's no hero and no villain.

It's clear they're in a different league when their pumps hit the ground and they start running. For most of us, the pump-hose test is a gradual unwinding; you start with everything in check, limbs in close to your body, posture straight, but eventually the limbs flail and the body doubles over. Not with Chris and Brad. The perfection is audible—nothing but the thump of feet, the rhythmic whisper of properly positioned arms and legs, the hard but measured breathing.

Brad is the winner; he does it in two minutes and two seconds. It's a new record for the Rangers.

With their efforts, Brad and Chris have ruined themselves for the rest of the day. When Brad finishes, he collapses on the spot. A while later he crawls over to one of the trucks and lies down completely underneath it to get out of the sun. Chris wanders around, no longer as present in the festivities.

My shadow in the grass is starting to look more like that of an average guy.

THE HOTTEST MONTH of the year has arrived and we haven't made one extra dollar yet.

Because July is also the halfway point in the season for many crew members, it's a natural spot to compare stats. One deployment is about average at this point, two are good. To the best of my knowledge, three have happened only once in the history of the Rangers, in 1998. Zero fire days by July 1 has happened before, but not for at least ten years.

Crews in other parts of the province aren't doing anything either, a comforting thought in the petty realm of inter-crew jealousies. Nobody is happy when one crew has more fire time than another crew. There's too much money at stake. On average, a unit crew member can expect to make about $5,000 in a two-week deployment. There are often situations where a unit crew can be two whole deployments ahead of another crew.

There's no great conspiracy at work, though. With thirty unit crews spread across the province, there are bound to be discrepancies. One crew that always seems to kill it is the Fort St. John Rhinos. (An odd name until somebody pointed out that rhinos stamp out fire. Even less odd when I found out that many Rhino crew members are rhino-like in stature.) Fire season comes early in northeastern BC, peaking around the summer solstice, when the near twenty-four hours of daylight dries out the forest. The Rhinos are also in oil country, so most fires are close to industry activity and need to be extinguished. Move a few hundred kilometres west to the Dease Lake area, and fires in similar terrain are left to burn because there's nothing around to protect.

July is the beginning of the end for the northern half of the province, but the southern half—Prince George and lower—is just starting up. Areas like the Okanagan and the Kootenays tend to start burning at this time. If it's a hot year every crew from the north, including the Rhinos with all their local fire time, will migrate to the south. We try to take comfort in the fact that the downside of being on those northeastern crews is that they have to live in the northeast, in unappealing industry towns. But they seem just as happy with their lot as crews from sexier locales like Kelowna and Nelson.

As it is we're still making base wages—$1,200 every two weeks. It's not heatstroke hot, but it is warming up and we're at that point in the summer when fires will burn even if it seems like they shouldn't. Lightning will pass over a lonely mountain in an evening squall and leave embers to smoulder overnight, waiting to be truly born in the sun's heat the next day. A spark from a bulldozer blade will shoot into the finest of old man's beard lying on the ground, and the operator won't know until he sees a blue wisp of smoke out his cab window later in the day. It's all liable to happen now.

IN THIS WARMING weather I drive to a chainsaw course with Dan. He's tired; I rarely see him like this. A crew supervisor and father of two young kids with a house in mid-renovation, he's working through that tough phase people hit in their late twenties and early thirties. Every day is long, every rest deserved. Despite his busy life, Dan still wants to be a performer, make people laugh and be the centre of attention. These traits are the difference between us—his personality is spectacular, like a stray tire let loose on a steep hill. I once saw him badger an acquaintance about why her relationship had recently ended until she was forced to explain that her husband had died in a car accident. That's the worst I can recall. On the other side, I've seen people he's taught at boot camp run into him on a fire later in their career, and their faces light up like they've waited years to be in his presence again. Still, spending time with Dan means spending time in his shadow, and it gets tiring.

Dan once told me he sometimes gets to a spot on a fire—always somewhere up high with a view—and he can picture himself and somebody else coming upon the spot on horseback. It's a vision of the past. He sees himself and the other person sitting there on their horses paused and looking out, not saying anything. The part he loves is how they don't talk. There's an understanding between them. They've ridden together for so many hours and are so attuned to each other's thoughts that an entire conversation passes in silence.

I think Dan has this vision because in real life he's not this guy. The cowboy is his ideal self, but his real self couldn't be on that ridge without some conversation or singing or doing something obnoxious and funny. Today, though, Dan would be silent at the imagined viewpoint.

I ask him how his dad is and he says not good—he recently had a stroke because of a lifetime of smoking and now can't see out of one eye. We talk about the family business. Both our dads run logging operations, his in the interior, mine on the coast. I say sometimes I want to go out and work for my dad. He does as well.

"When you're young, you want to get out and do your own thing," he says. "You want to turn away from that lineage, but as you get older you want to be part of it. You think maybe it's not so bad doing that."

I feast on this comforting sentiment. Day-to-day work is a grind; seasonal work can suck. It's nice to imagine being settled. I think that's what people want—they want to be settled and at the peak of their working powers. I'm not there, and working in Dad's shadow isn't the answer. It's just a pleasant, useless thought.

ABOUT TWENTY OF us are attending this three-day chainsaw course. Crews from all over the Northwest Fire Centre are taking part—they come from Burns Lake, Terrace and Hazelton. There are members of other twenty-person crews like ours, as well as members of smaller three-person Initial Attack (IA) crews. Before the course starts there are formal introductions around the room—Where are you from? How much chainsaw experience do you have? We're friendly with each other, but there are touches of underlying resentment born of boredom and young egos. We find things to pick apart about other crews: whether they dress in uniform, how fit they are, how organized.

The best way to improve our chainsaw program is to have people go out and actually use chainsaws. Instead, we theorize in a classroom about best practices. A snippet of the conversation goes like this.

"So, if we have a crew in the bush and the supervisor is also running the saw, who supervises the supervisor?"

"We don't need someone to supervise the supervisor every time. That's why they're a supervisor."

The goal of the bureaucrats is to eliminate grey areas that can never be eliminated. One of the instructors sums up these problems and, inadvertently, the anxieties of the entire bureaucracy when he says, "I hate to say it, but sometimes you have to go with your gut."

ON THE LAST day of the saw course, we go out into the field. Dan stayed with a friend in Houston the night before, so it's just Warren, Rob and me driving out in the morning. We're going to a training site south of Houston near where I did my falling test with Brendan more than a month ago. We have to be at the training site early and I cut it too close; I don't give Dan the punctuality he needs. People from other crews will be there before us. We'll arrive at the same time as the colleagues we're trying to upstage.

I can tell Dan isn't happy when I pull into the parking area, and we don't look at each other. He has the ability to express major disappointment with a minute shift in body language.

After a meeting, we go into the bush on our own and start falling trees. I'm still irritated at Dan and I work fast. My stumps start adding up, and by the end of the day I have forty or fifty trees on the ground. All the destruction softens my anger.

Two trucks are going back to Telkwa at the end of the day. It's going to be Dan and me in one truck and Rob and Warren in the other. All of us know this will be the arrangement long before the end of the day; Warren and Rob have sensed my infraction. In a job where privacy is hard to come by, two hours one-on-one is more than enough time to talk things over.

I climb in and waste no time getting to the subject of the morning's lateness. Dan gives the spiel on how we need to be better, we need to be an example to others. I don't disagree. Perhaps this episode is only frustrating because I know the problems, and we pride ourselves on being an organized crew. I've come a ways in living up to this standard, but I'm still nothing like Dan. Dan would create a spreadsheet to track his peanut butter use. I'd lose half of mine on my own shirt. We speak openly about the lateness and it allows us to move on in a hurry.

OUTSIDE THE CAB of the truck, the foliage inches up from the ditches. We drive by a river filling up with salmon; we see the snow receding from the mountains, their newly exposed brown flanks looking alien in this land-scape defined by winter. Summer saturates the land like an ocean tide, pausing for a minute at high slack before receding as quickly as it came.

As the world comes alive, our impatience for fire season grows. We're bored with all the training and project work and there's a seed of agitation germinating in the crew—especially with the veterans, who have just been through two seasons of very low fire activity. I need to find hope somewhere, so I start looking at the weather. At work we have access to all the technical forecasting. There's the Fine Fuel Moisture Code (FFMC), which measures the dryness of small debris such as needles and branches in the woods. Another is the Drought Code

(DC), measuring how dry the forest is at a greater depth, indicating long-term drying. The FFMC numbers can fluctuate significantly in a matter of hours; the DC rises and falls at a much slower rate. Five millimetres of rain will send the FFMC rating from high to low but will have virtually no effect on the Drought Code. The most tantalizing weather check, though, is to scroll through the various online weather maps and fire reports available for anyone to see at sites like bcwildfire.ca and ciffc.ca (a nationwide fire information website). At CIFFC there are stats showing anything from how many crews across the country are working on fires to how many rolls of hose have been sent between provinces to aid in fire-fighting efforts. The BC wildfire site has maps showing precipitation, fire danger and temperatures. The maps are coloured with big splotches of green and yellow and sometimes orange and red. The little icon of a cartoon sun, the furious clots of lightning bolts, the red pool of a dry precipitation map. It all becomes a digital altar.

FOR PROJECT WORK, it doesn't get better than falling. But even the best project work becomes a drag as we poke further into summer with no scent of real fire.

I'm partnered with Warren for a day of falling practice and at first we take our time putting perfect cuts into each tree we fall. But the trees are small and non-threatening, making it hard to focus. I get reckless as I burn through my tank of gas, pushing trees with my shoulder when they don't crack off the stump right away, forgetting to look for brush falling from the treetops.

I look back at Warren to see if he's still clear of my falling area and he's standing oddly still. His head is tilted down at his radio, which is sitting in a chest pack that he's holding up closer to his face. I shut down my saw and turn up my own radio. I hear Dan conversing formally with one of the dispatchers at the fire centre. I tense up, completely immobile as I listen to the instructions.

Fake work is over. We're going to fight a real fire.

2
First Fire
July 2014

On our drive back to the base we see smoke from a fire in the distance. The smoke is low to the ground, sitting on its haunches, waiting for gusts of wind to send it stampeding across a hayfield.

At the base, people get their gear ready fast. We change into our uniforms: red button-up shirts and blue pants made of a fire-resistant material called Nomex. It's tense. Packing up feels like frantically memorizing your last lines before going on stage. People stand by their trucks with a distant look in their eyes, mouthing a final count of hose, or they stare hard into their lockers before unloading a question on whomever is closest: "Do you have bug dope?" Feet stomp in and out of the locker room, which is separated from the trucks outside by a large, open bay door. Gear is heaved into truck boxes, where waiting hands tie down the loads.

Once packed, we're told to wait. There are a couple of fires burning in the area and the fire centre needs to decide which one to send us to, or if it's even necessary to have all twenty of us go. The locker room takes on the sounds of tools being sharpened and bits of conversation. The sounds of waiting.

After half an hour of this, half the crew is sent to a fire a little more than an hour's drive from the base. Somebody several steps above Dan at the fire centre office has made the call. Warren will be leading this ten-pack of Rangers. In losing half his crew, Dan's mood becomes foul. Nobody wants to go near him. Before Warren's group leaves, Dan takes him aside and they talk quietly in front of the locker room. It's breezy outside and the bay door makes a massive frame in which Dan and Warren look small. I stand nearby and listen. There's a fierceness in this moment.

"Don't let them push you around. Use your head," says Dan.

"Them" is fire centre management. Now that we've been separated, Dan is worried it will stay that way. Warren's group could be sent out to a remote fire for a few days, where they would be "camping on the line"— camping somewhere near the fire without any amenities. We keep three days' worth of food and water in our trucks for these situations.

Splitting up the crew isn't ideal, and Dan takes serious issue if it's even mentioned as a possibility. I'm with him on this one—overhead underestimates the value of keeping us together.

Warren's group of ten rolls out of the base and the edgy feel of waiting subsides. With fewer people, the atmosphere is more casual.

Less than an hour later we're told to go to the same

fire. I know Dan would like to pick up the nearest phone and call somebody for an "I told you so." We settle for a quick eye roll before jumping in the trucks.

Most of my squad rides in Charlie truck, one of five full-sized crew cab trucks leased for each unit crew every summer: one truck for each of the three squads plus an Emergency Transport Vehicle (ETV) and a truck designated as the Ranger One truck (Dan's call sign).

We get these trucks brand new, and the wildfire decals on the side, the growling diesel motor and the high-elevation seats make for potent sensations of invincibility and coolness.

In the back of Charlie truck are Kara and Brian. Tabes and I are up front. Tabes is Graeme Tabor. Bearded and stoic, he looks like he's walked out of a photo from the Klondike gold rush.

On the floor between the front seats is a VHF radio, used for serious communication on fires but used for jokes and harassment when we're driving.

Other personal effects are stored here and there—magazines, sunscreen, lunch containers. We haven't done any firefighting yet, but already a stiff grime coats the interior of every truck, making everything feel like tacky snot.

It's two in the afternoon on July 2, and we're heading to our first fire. Most years we're out the door within a month of arriving on base, sometimes earlier. One year we were called to grass fires up north on our second day. Today is the drawing of a line between two realities, the time before your first fire day and the time after. It feels like day one on a new job. It is day one on a new job.

Our access is up an old logging road with deep water bars slashed across it every thirty metres or so. Water bars are trenches running across a road to help with drainage; they're dug out once a road is no longer frequented by industry. Each time we hit a dip I brace myself, thinking the nose of the truck will slam into the dirt like a javelin, but each time we narrowly avoid it, sometimes scraping the back bumper in the dirt on our way out.

We see smoke from the highway, but we lose it as we get deeper into the forest. These are wild woods we're heading to, deep green and stretched taut across the landscape. There's nothing but logging scars until the ocean. We catch a glimpse of the smoke above us a couple of times, but the road is a tunnel through the forest and more often than not we're nose-diving into a water bar.

We come around a corner and a view opens up as we start hugging the side of a small canyon. Directly across the canyon is the fire. It's not too big, maybe ten hectares, a solid six or seven Walmart parking lots, but it's active. The smoke surges from the ground, a huge stream of black and grey that's tinged yellow and hurtling into the sky so quickly it looks like somebody's pressed fast forward on the scene. The wind coming up the valley pushes the fire hard against a plot of young, planted trees about the height of a basketball hoop.

Started by lightning, this fire is on the same site as an old burn from about twenty years ago. In a healthier forest it would be a canopy or crown fire, burning from the ground all the way to the tops of the trees. As it is, the fuel it's burning is too sparse and too young to allow it to really get moving.

As we linger around the trucks getting ready, I see looks of awe on our rookies' faces. Relative to other fires this is tame, but there are enough flames, smoke, crashing trees, and bucketing helicopters for them to be stealing glances. There's even a bit of fire noise, like the sound of a roaring creek heard from a distance.

We start cutting a hose trail toward the fire. We won't attack it directly—that is, we won't work at the front of the fire, where the wind is pushing it into the plantation—instead, we'll cut trail from the creek at the bottom of the canyon to the back edge of the fire. From there, we'll set up pumps in the creek and hook up hose until we're at the fire's edge.

To cut the trail we split into several groups of two. One person runs the chainsaw, cutting a swath about a metre and a half wide, wide enough for a stretcher in case of an injury. The other person cleans up behind the cutter, normally a job for the less experienced. We call these pairings cutters and swampers. It isn't glorious work for either party, spending their day bending and standing. Cutters need to be nimble with their twelve-kilogram saw and swampers are faced with the menial task of picking up sticks next to a screaming chainsaw all day.

Dan, Kara and I go ahead and lay ribbon; cutters will follow a trail of flagging tape tied to tree branches showing where we want them to cut. Since Kara is only in her second year, this is her first time at a task that requires planning and judgment. Dan explains to Kara how to do our job faster by having one person hang ribbons farther apart and the other come in behind to fill in the gaps.

The three of us tumble our way to the bottom of the ravine that separates us from the fire, laying ribbon as we go. There's devil's club everywhere, mosquitoes swarm our faces, and we trip over unseen roots and step through rotten logs. It takes patience to walk through ground like this. There's no animal shit in this part of the woods; only squirrels and birds are small enough to find a home here without being ripped to shreds.

We reach the fire and the sound is louder, a giant, steady gasping for air. Everything around us burns with fury—the brush, the dead trees on the ground, the stumps. Most spectacular are the standing dead trees from the previous fire, which are burning again and covered in honeycomb-like embers, sometimes from top to bottom. Occasionally one of these trees will snap off in a gust of wind, and the stem will glow dark orange against the blue sky for a second before shattering on the ground and being consumed by the surface fire.

Kara's new to this, standing next to a fire, sizing it up. Waiting and thinking. I've had this moment many times with Dan. For the better part of a decade we've stood at the edge of uncontrolled fires, partly strategizing, partly in awe of the moment. Kara is restless; when she glances at us it's always a question but nothing is verbalized. In the background the sound of half a dozen chainsaws fills the ravine as the crew gnaws its way through the bush, following our ribbon until it meets the fire.

We start losing light as the first cutting groups arrive at the fire's edge. They come out of the bottom of the ravine with sweat pulsing from their bodies. The veterans read the situation and take a minute to clean up their gear and watch the burn. The rookies stand, uneasy. We gather up and drive home in the long dusk. I follow the tail lights of the other trucks and I'm soothed. Finally, a little fix. It's eleven p.m. when we finish work, which means we can't start again until seven a.m. the next day. We need a minimum eight-hour break between shifts. Eight hours is whittled down to five when you factor in commuting and eating.

THE FIRE IS quiet the next morning, slowed by calmer winds and the high humidity that descends on forests in this part of the world. It has also been completely circled by a bulldozer that worked through the night, accessing the fire near the tree plantation. We call the wide strip of exposed dirt left behind by the bulldozer a cat guard, a guard for short. The cat guard is dug deep enough to get below the organics of the forest floor and down to fire-stopping mineral soil. It's different from the fuel frees we cut with our chainsaws. Cat guards are wider and they expose much more dirt than we do with our skinny hand guards. They're also hard on the land and messy, but it's a low-cost and relatively effective way to stop fire.

From our end, it's a treat working off cat guard. We do so much struggling through the bush that walking a guard feels like being on one of those human conveyor belts at an airport.

As I'm getting set up for the day, Rob walks by directing an excavator. It's being used to knock down trees close to the guard, making it less risky for us to hose the fire's edge. Even though it's Rob's fourth year firefighting, he's never worked with a machine before; we've had a lot of quiet years in a row.

Rob grew up in Toronto and his high school had a bigger population than most communities in north-western BC. He sometimes tells stories about growing up in the biggest city in Canada, a place most on the Rangers have never even visited. The biggest laughs he gets are when he re-enacts scenes from his high school dances wherein the girls did a lot of twerking and the boys tried to play it off like that sort of thing happened all the time.

After high school Rob moved out to BC and went to university in Prince George. He did an outdoor education program and worked as a tree planter for several years before he started firefighting.

Rob watches as the excavator, with minimal effort, pushes over big crusty trees that hit the ground and rattle his boots. It's tempting to yelp and cheer while watching this, but by the code of the bush, Rob stays casual, much like the boys of his high school dances.

I spend most of my day walking somewhere between ten and fifteen kilometres, up the cat guard to where everyone's hosing and back down to the pumps. The walking is one of the big differences between being a crew member and being a squad boss. You have to see what's ahead for the rest of your squad and, to a lesser degree, the rest of the crew. Walking is a big change, but being in a position of authority over your peers is by far the toughest change. To spend years at a job where your co-workers double as your closest friends—not just "work friends" but friends in the deepest sense of the word— and then step even the slightest bit away is almost not worth the step. I remember talking to a friend outside the crew in my second year of lower middle management. A little drunk, he told me that the word from the rest of the Rangers was that I'd lost my way—I was just a shill for Dan. I knew this kind of talk was happening, but man did it hurt.

Those feelings go away, though. The position gets more comfortable every year, and the only hurt I'm facing now is that I'm doing all this reconnaissance in new boots and my soft toes ache under the strain of callus formation.

I run into Dan on the guard and we have a minute to talk. We haven't been alone since the crew was almost split up yesterday, and I start venting about the bureaucracy and how it chokes our ability to do work. He opens up about his own frustrations with it. I ask him if he thinks things are getting better. He doesn't often stall, but he hesitates on this question. He says you carve out a niche in the world where you have some control over things; you make it your own as much as possible. He says maybe we're lucky to sometimes battle this incompetence. It's made us shine brighter when perhaps we were nothing special to start with.

The rest of the day passes uneventfully, hosing the edge of the fire, now terminally ill.

ON THE THIRD day my squad is paired with a three-person Initial Attack crew from Houston. Dan has been badgering their leader during this fire. Any of us can take it, but Dan's brash manner with these strangers is uncomfortable.

The leader of the IA group offers to have one of his guys look after the pumps for the day. I say yes even though I'm not confident in their abilities. I'll make sure I have an eye on the pumps as well.

I walk by Dan after the meeting and he confronts me about it.

"Did you put them in charge of water?" he asks.

"Yeah."

"That's fucked," he says.

It's trivial in my eyes and anger engulfs me as I shrug

him off and walk down the cat guard to where we're hosing. Dan can have no tact. He'll fight any battle.

I get to the line and start setting up for hosing, hoping time will calm me down. Coming off the main hose lay, made up of inch-and-a-half-thick hose, are little faucet-like pieces of hardware we call "thieves." Here we thread on what we call "econo"—fifteen-metre chunks of hose the same size as what you'd use on your lawn. The fire has quieted down so much that there's not a whole lot of hosing left to do. Bigger logs have retained their heat, same with stumps, but we can cover land pretty quickly; once we're done with a section we leapfrog down the hose lay and set up our econo at the next available thief.

Dan comes up the guard to where I'm hosing. I avoid talking to him. He's falling trees with Blaine, who joined the crew in 2009 as part of the Ministry's Youth Employment Program (YEP). There's always debate as to whether your YEP year counts as a season on the crew. In Blaine's case, he spent most of July and August on the fire line and it's hard to discount that experience. He spent his eighteenth birthday working a big fire near Kelowna. We didn't really know him at the time but invited him for a swim in the lake after work that night. Inclusion was probably the best birthday present a YEP could get from the Rangers. Later in the summer we started calling him Higgins because according to everyone, "He just looked like a Higgins." He was baby-faced and had curly blond hair but with the deep voice and mannerisms of somebody much older. The next year on his birthday we bought him a cake from Dairy Queen. On it was an image of two black socks and the words "Happy 40th, Higgins!"

An hour or so later, Dan and I are both eating slices

of pizza. From a little ways away he raises his slice to me. Perhaps holding up a slice from a distance is his apology. I return the gesture, whatever it means.

By mid-afternoon, hosing is done and we break into a patrol, walking along the edge of the fire looking for leftover bits of heat. Patrol is the last stage of firefighting, boring but necessary. My patrol group consists only of my squad.

Dan, Warren, Rob and I had a meeting to choose the squads back in May. Each of us got to pick a member based on our seniority; I chose first, then Warren, then Rob. Each squad needs a designated faller as well as a designated first-aider. Tabes is my squad's faller and a rookie named Bronson is our first-aider. Other than that, the picks are largely random. Rounding out Charlie squad are Kara, Brad and Brian. So far, I've spent the least time with Bronson, and I discover on this fire that he's fond of pushing people's buttons. He's kept in line by Brad, whose take-no-bullshit attitude is backed up by the fact that he will actually get into fights with people he sees as a threat to the balance and well-being of whatever social organization he's part of. However, Brad's duty as Bronson-tamer will be cut short, as Bronson will break a bone in his foot not long after we finish up on this fire. Bronson won't be replaced and our crew will run at what Dan calls a "lean nineteen."

Squad cohesion is only really tested when we're out on a fire. When we're on base and doing project work, Dan has the entire unit crew working together as much as possible.

My anger lingers from the morning, but the patrol neutralizes the poison. I have a great squad this year. I can tell this within five minutes of starting the patrol. There are enough jocks and enough geeks to make it work. Too many jocks and it would be one long fight for control—a four-month battle for dominance. Too many geeks and nothing would get done without an explanation, and after the explanation they'd be liable to waste time thinking of different ways to do their job. We're evenly matched this year—jock and geek, doer and thinker, Type A and Type B.

Suddenly I feel lucky, lucky to have this job and be outside working with these people. We circle the small fire a couple of times. This thing is out. Some buckets from a helicopter, a bulldozer and twenty people have silenced its brief roar.

WE'RE DONE WITH fighting fire for now but we'll be put on standby for the weekend. On standby we get paid a third of our wage from 9:30 a.m. to 9:30 p.m. In exchange we must be within a half-hour of the base and sober. The curfews and boundaries make standby feel a bit like being a teenager again—lots of freedom, but ultimately still living under someone else's rule.

The first couple of weekends of standby are always good. It brings the crew together. We tend to hang out more, as we're all equally stuck. Because we're deprived of booze until 9:30 p.m., people also tend to drink harder on standby nights.

But long periods of standby are wearing. This is especially true in the Northwest, where urban amenities are minimal and cellphone coverage is so poor that even the lakes and trails near town are off limits. It can feel like prison.

In 2008, a quiet summer with long stretches of

standby, I decided to break out of prison. While under the rules of standby, a crew mate and I drove four hours to Prince Rupert to hang out for the day. Maybe *decided* is the wrong word, as this was the kind of adolescent-bonehead move that's less a decision and more a complete lack of judgment. And I wasn't even a teenager, I was twenty-three. Looking back, I think it was a turning point in my life, a final instance of whatever it is that causes our young brains to be so erratic.

Obviously, the crew was deployed while we were in Prince Rupert and we returned to Smithers to spend a week on the base while the crew was out fighting fire. We were lectured by upper management but the only effective (and worst possible) punishment was the crew being deployed without us.

A FIRE HAS started near Tumbler Ridge in the northeast and has quickly grown to thirty-five hundred hectares. Crews are being sent there from all over the province. I'm running through scenarios in my head. Who else will they send? Will our fire centre let us go anywhere?

Tabes and I take a trip to town to pick up chainsaws we'd taken to a shop for repairs. Similar to Rob and many before him, Tabes comes to the crew from Ontario. Young outdoor types from Ontario love to come out to BC to fight fires or tree plant. This is met with trepidation from some people in BC. When it comes to these jobs and the "BC lifestyle"—living in resort towns to maximize time skiing, biking and climbing—the enthusiasm of these easterners can be grating.

Tabes isn't that kind of Ontario guy, though. He's from Thunder Bay, the original gateway to the Canadian west. This makes him a sort of honorary member of the region. Tabes was obsessed with downhill skiing as a kid and came to live in the mountains as soon as he finished high school.

When Tabes and I pull into the parking lot of the saw shop we see Marc. He works at the fire centre and lives on a hobby farm east of town. He's quiet and fit. When not in his work uniform he wears gumboots and a Tilley hat. But today he's in his work uniform—blue pants, a blue T-shirt, running shoes and sunglasses.

I roll down my window and hot air rushes into our cool truck.

"What's happening?" I call out the window to Marc.

He switches direction from the saw shop to our truck.

"On my way to Tumbler Ridge," he says. He's a pretty reserved guy, but he betrays some excitement. Tumbler Ridge is the greatest place on earth right now.

"No way!" Tabes says from the passenger side.

As cool as possible (not cool at all), Tabes and I squeeze every bit of info we can out of Marc. How many crews are going? How big is it now? Will there be a fire camp?

Marc is what they call an "ignitions specialist." I think he used to work for Parks Canada. A "parks guy" is what we call these people. I don't know what this means exactly; it's not a stigma but it's not necessarily admirable either. Ex–parks guys are a strange bunch. All of them seem to be from back east. They're quiet and they're quietly bonkers for all things nature. They've tried every type of outdoor activity. They have all the gear—Gore-Tex everything. Among the mob of easterners coming to BC to stake a claim in the extreme sports gold rush, the parks guys are high-status. They've climbed the mountain and they sit in solitude, looking down at their eastern peers labouring in ski hill bars and on gear shop floors.

Anyway, I'm not sure if Marc is an ex–parks guy. I'm not even sure he's from back east. But Marc is on his way to Tumbler Ridge, pulling a trailer of burn fuel to be used for aerial burn-offs. A burn-off is a fire that's deliberately set in front of a forest fire to try to reduce available fuel. Aerial burn-offs are rare and employ a massive torch dangling from the bottom of a helicopter, and you can imagine the kind of machismo that surrounds this tactic. If this is the high point of Marc's life, he's had a pretty good life.

In the afternoon, I hear about another fire blasting off south of Vanderhoof. This charges me up even more. Things haven't looked this busy in years. But I'm one of the few who understands the potential. Warren, Dan, Tom and one or two others have been here before. But the faces of our newer guys are blank. They've never sat on this invisible precipice.

AFTER WORK A bunch of us jump off the Telkwa River bridge. Brad is eager to climb to the top, reached by scaling a series of old, warped spikes pounded into the creosote beams. In the middle of the climb I nearly lose my nerve. All female members of the Rangers are down on the riverbank watching, a spectacle that used to spur me on; today it doesn't cross my mind. Instead of being boosted up the bridge by testosterone and fit women in swimsuits, I'm nudged along by fear of losing all dignity if I turn back.

When we get to the top, the only thing that tells me

this is okay is knowing that I've done it before. After I jump, Brad does a front flip. He hasn't jumped off this twelve-metre-high bridge in over a year and he's able to throw himself upside down over top of the river. He's stretched out for a second in the air, muscles taut and focused on maintaining control. He lands feet first in the water.

FIRES ARE BURNING across the northern half of the province, but we're still at the base.

On our second standby weekend I go swimming at one of the only lakes with cell reception. I swim out for ten minutes at a time, then come back to check my phone. Beyond the lake's far shore I see a light-brown smog brushed over the blue sky. Another fire has started burning, this one deep in a remote provincial park and left alone by the Ministry.

I come back and check my phone again. No missed calls.

3
Euchiniko Lake Fire Tour
July 13–27, 2014

'm walking across the front lawn toward my house when Dan calls and says to head to the base. First, I have to call all the members on my squad. I try to keep my cool when I call the squad but it's hard, and my words end up coming out all breathy and excited. I get my gear ready and Dan calls again.

"Stand down until tomorrow morning," he says.

I call everyone again and pass on the message.

Dan calls a third time. "Tell them to pack a lunch."

"Okay."

This one is a no-brainer. We bring a lunch to the base every day, even if we're going on the road.

I call everyone a third time. The rookies appreciate my lunch reminder. The vets say, "Thanks, Mom."

I don't sleep much that night. I never do when we're on our way to a fire. Going to Ontario or Quebec is the worst for not sleeping. The time change blows up your already fragile pre-fire sleep patterns. Two years ago in

Ontario I stayed up all night. We were on our way to a fire and staying at an unused summer camp—a bunch of rusted-out trailers and huts in the woods.

Forty of us from two different unit crews tented in a nearby field. I was nearly asleep when, at about eleven p.m., the caretaker came around. He was softly calling "Pepperettes, pillows" as he walked between the tents. He was drunk. In one hand he had a few moth-eaten pillows, in the other he had a handful of pepperoni sticks. As his voice faded into the night, the laughter started moving between the tents like signal fires. I was up for good.

IN THE MORNING I show up at the base at 5:10 and I'm late. Not late on paper—we're to leave at 5:30—but everyone has done everything. Trucks are packed and uniforms are on as I walk across the parking lot and get my things arranged.

We're going east to Vanderhoof, another northern town powered by pine trees and their transformation into two-by-fours. Once there we'll turn south and drive 120 kilometres on logging roads to where a fire is burning aggressively through a mass of valuable timber. It's also close to an expensive fishing lodge and other backwoods cabins. It's known as the Euchiniko Lake fire.

Charlie truck has been our truck for a couple of months now, but starting today it becomes Our Truck. The transformation is due to this being our first four-teen-day deployment—the maximum number of days we can work before having to take three rest days. It's also because we'll start the deployment—the "tour"—with a five-hour road trip and do plenty of commuting once we're out working the fire.

Truck dynamics are perhaps the biggest factor in our mental health when firefighting. Having a good relationship with those you drive with can mean the difference between a good season and a bad one. We enter the trucks at our most tired and defenceless moments—before the workday begins and after it's over. Because we don't do actual work in the trucks, the bonds formed inside their increasingly filthy interiors tend to be more familial than work-related. Truck mates come to stand in for siblings, parents and spouses. They also offer rare solace from the social demands of a twenty-person crew that operates, for the most part, without a solid wall of separation.

Charlie truck is off to a good start. It's important to have a rookie to tease, which we do. Only two of us are interested in driving, an ideal number; one isn't enough and three is too many. We'll also be spared competition for the front seat: Brian and Kara, as rookie and second-year, respectively, have permanent back-seat status.

In Vanderhoof, the crew meets at the Co-op parking lot. It's high summer, so the light and the heat are of a midday quality, even though it's only eight a.m. Because it's a Sunday the parking lot is empty and the town is dead. The Co-op isn't quite open so everyone goes to 7-Eleven to buy chew and treats. After this we hang around the trucks eating bags of chips the size of our torsos, loitering like guests in the courtyard of a grand hotel.

Vanderhoof is a glorified work camp. You can buy the necessities for work and life and nothing else. The Co-op is the biggest store in town; in second place is the store that sells work clothes. It's a three-floored fortress of thick wools, hardy boots and country radio. The Co-op and the work gear store are separated by train tracks plied by freight trains that career through the middle of town, heading for the coast with treasure from the Interior.

The Co-op opens and we rush in with the fervour of looters, before we drive out to the fire.

On the way we pass rural farms, a lake where locals recreate and a small First Nations reserve. Cell service fades out and everyone throws out their final Hail Mary text messages before we're away from civilization for good.

THE PAVEMENT ENDS and we're on a network of roads used mainly by industry and ambitious rednecks looking for somewhere remote to shoot guns at empty beer cans. After half an hour, conversation stops. Caffeine and nicotine wear off and the drive becomes arduous. I struggle to keep my eyes open.

One of the biggest letdowns in my first year of firefighting was how anticlimactic it was to arrive at fires. We don't swoop into the head of the fire and leap from the trucks with hard hats askew, sprinting with pumps and hose to a nearby pond. Instead, we creep up on it in an orderly manner, like we're in line to get our passports renewed.

We arrive at the staging area for the fire—a bulldozed piece of earth big enough to serve as the foundation for a shopping mall. This staging area doubles as a safety zone, a spot close to the fire that has been stripped of any fuel and is big enough that, even if you were completely surrounded by fire, you would be, if not comfortable, at least safe.

It's hard to be patient, but there are signs of a good fire here—confusion, heavy equipment, helicopters, an actual fire. Sometimes we arrive when the fire is nearly out and all we do is a few days of patrol along a mostly cold edge.

Dan and I take a flight in a 206 helicopter, an older-model chopper with interior upholstery straight out of the '70s. Fires this big are usually assigned at least a half-dozen helicopters, with one or two dedicated to shuttling crews around. From the air, the ground looks used and worn out. There are cutblocks—areas that have been logged—everywhere, enough to make them look like a natural feature of the landscape. Between the cutblocks are patches of virgin timber tainted with pine that's been gnawed to death by the pine beetles that have infested this region for twenty years. To the west, in the very far distance, is the tail end of the Coast Mountains. Everything else is interior plateau, a place we've been relentlessly exploiting for a century. It doesn't have the majesty of coastal timber or snowy mountains. In the absence of these factors the Interior of BC has been tilled from top to bottom—logged, mined, flooded by dams. Humanity has taken this land by the hair and dragged it around.

We'll be heading downhill toward Euchiniko Lake, assisting equipment in building a guard that will hope-fully stop the fire. Looking at land to be worked from the air can be deceiving, though; from a hundred and fifty metres up, what appears to be a grassy meadow can actu-ally be head-height brush. The chopper sets down and Dan goes to talk more tactics with other supervisors.

I join the rest of the crew in the safety zone. Crew members keep walking up and asking me questions. My replies are short. "How's it look?" "Pretty big." "Is there a plan?" "Sort of." "What time are we hitting the line?" "Maybe never." The terse replies are an attempt to curb the excitement, already too high on account of all the new people.

In the late afternoon, Tabes and I are put to work doing some line locating. It's similar to what Kara, Dan and I did for the crew on our first fire near Houston, only this time it will be for bulldozers pushing guard. We drive down a logging road and park the truck at a fork in the road. After the sterile, cold air coming from the truck's air conditioner, the outside world wallops me. It's not so much the smoke and heat as the sick feeling that everything around me is cooking. For a few seconds, my reptilian brain finds all this wrong and threatening.

We put on our hard hats and grab our backpacks and Pulaskis—the bare minimum we carry with us at all times—and walk down the road into a wall of smoke. The flames are still hidden by the hill sloping down in front of us. After fifty steps I'm jittery and soaked in sweat. I feel pressure. The decisions I'm making matter. "They're just trees," they say. Yes, but it's also pride and timber supply.

We leave the road and cross a cutblock, hanging ribbons as we go. The late afternoon heat smothers us. There are only a few weeks each year when heat like this is possible in this part of the province. Each of those days is shocking in its intensity.

When I scuff my boot, the soil rises in a cloud of brown dust. There's no moisture to glue anything to the ground. We come to the bottom of a hill that spills out onto a marshy area. The fire is burning down the hill in old-growth timber. Seasoned beetle-killed wood has draped itself all over the still-living timber, acting as a sort of nutritional supplement for the fire. There's a ranking system for wildfires used by the BC government. Rank one is the lowest, a quiet, smouldering ground fire. Rank six is the highest, its descriptors being "conflagration" and "blow up." This is fire so powerful it can literally rip trees out of the ground. We're dealing with a rank four here, a steady flame front on the ground with occasional "aerial bursts through the forest canopy."

We gauge how close to get, deciding to hang ribbons about fifteen metres away from the flames in anticipation of more burning before the bulldozers arrive. We hang ribbon for another hundred metres across the marsh but then the fire picks up. From this distance, we can see the smoke getting dark and thick on the hillside. A few of the ribbons we hung are about to melt in the heat. We scoot back and reroute our line.

With the reroute we start tracking far away from the fire. We work through more cutblocks, bits of old-growth, swamps. We're following the bottom of a small basin. The fire burns above us, flames curling off the treetops and into the sky. If we can keep this thing corralled so that it's burning downhill into wetter ground rather than up a dry hillside we'll be okay. Tabes and I go hard with the line locating. Soon I'm in a rhythm: eat two Twizzlers, drink water, beat through the bush with ribbon. Repeat every half-hour.

After thrashing through a section of bush we find a steep outcrop. From here we can see the rest of the western edge of the fire burning down toward Euchiniko Lake. Black smoke pours out of the trees about a hundred metres away from us. Our view of this nearby flare-up is mostly blocked by the broad-leafed brush we've been fighting, but any gap in the foliage is coloured bright orange by the wall of flame. Down the hill the fire burns with a range of intensities depending on what type of fuel it's in—brush, dead pine, living spruce, some mixture of the three. We hear the soft buzz of crickets and the intermittent thump of distant choppers, but the most prominent thing we hear is the hair-raising sound of fire.

"It's loud" is all I manage to say about the scene to Tabes, who simply nods in agreement.

The sun melts into a thick pool of low-lying smoke. Tabes and I take a minute at our viewpoint to gape at the result of a timely lightning strike finding dry sticks. I'm beat. The Twizzlers aren't doing their job, but Twizzlers aren't usually asked to do this much. Dan calls us back to the trucks and the workday is over.

CAMP TONIGHT IS a bug-infested rec site with one decaying outhouse. It sits at the edge of a small lake with a ring of reeds around it and a few ducks paddling out in the middle. These rec sites are installed by the government and there are hundreds of them in BC. One of the squads has come back early to set up. A row of ten blue tents sits in a knotty patch of grass just up from the lakeshore.

It's getting dark, but we still have to make dinner. I shred a mountain of Co-op brand chicken breasts with my toothpick-sized knife while others cut vegetables and boil rice. Dinner for twenty is made without a countertop, on a two-burner camp cookstove. The rookies, who stayed in the trucks all day, are restless. They gallop off to collect firewood and get their things in order for the next day.

My tent partner is the high school kid we hired. His

name is Jeremy, but we call him "the YEP." He's got that teenager skinniness that will probably fade in a year or two. His expression borders on a smirk, which may just be his way of coping with the overload of new people and a new job. I don't know much about him beyond the fact that he hasn't endeared himself to anyone. Each person on the unit crew has an opinion—how to fight fire, where we should camp, which way we should slice chicken breasts. To keep the garbage pile of opinions manageable, there's a culture on the crew that the newer you are, the less your voice should be heard. Jeremy hasn't caught on to this yet.

He keeps asking me questions as we're going to bed—what to do about our unsecured tent fly, what's going to happen tomorrow. I give him short answers and never once look him in the eye. I feel bad, but I have nothing to give him.

IN THE MORNING, mosquitoes are crowded between the fly and the tent wall. It must be like an opium den for them up there, all that heat and human smell. I'm jealous of how easy their day will be.

I'm the first one out of the tents and there's steam coming off the lake and a loon calling.

Breakfast is quick and sloppy. There's yogurt and granola, bruised fruit, Froot Loops. The worst of the morning people blindly grab a power bar from the grocery boxes before slumping into the trucks. We drive to the staging area, leaving trails of the finest dust dyed orange in the morning sun.

Tabes, Brad and I are working with equipment today. We've moved farther down from where Tabes and I were working yesterday. Our start point is directly below last night's viewpoint. We meet up with the equipment operators and resume laying ribbon for them to follow.

Two helicopters are putting buckets of water on burning trees right next to where the machinery is putting in guard. It's an expensive battle and I don't know if there's ever been a more content bunch of equipment operators. Logging in the twenty-first century, especially in this region, isn't much different than being a worker on an assembly line. Everything is done with computer assistance from the cabs of the machines. But putting in a fire guard is dynamic, and with today's heat, there's sure to be excitement. When I stop to talk to one of the operators, he stands up out of his cab and leans down toward me. He's overweight and has on a stained ball cap and loosely tied steel-toe work boots.

"I feel like I'm in a fuckin' action movie!" he says.

As the machinery makes its way down to Euchiniko Lake, Tabes and I look ahead for the Alexander Mackenzie Grease Trail. The trail is an old trading route carved by First Nations traders. The grease in the name refers to the grease of fish, mainly oolichans, that were packed inland from the coast for trade. It's a National Historic Site and a First Nations heritage site. Settlers and First Nations are united in not allowing anything to disturb this trail.

In 1793 Scottish explorer Alexander Mackenzie ended up taking the grease trail during his search for the Northwest Passage. This was a sort of national pastime for white people for the first few hundred years after they showed up in the New World. Every year somebody was trying to find the Northwest Passage, and every year they failed at best and died at worst.

Important for us today is to stop the heavy equipment demolition party before it lays down a twenty-foot-wide fire guard over the Grease Trail. Given how fired up these operators are to blast unfettered through the bush, we may have to tie ourselves to the trees once we find the trail.

We're having trouble, though. Nobody knows the exact location of the trail. Tabes and I are convinced it runs right along Euchiniko Lake. Our logic for this claim is: "It sure is nice to walk along a lake."

When we discover a trail just up from the lakeshore I'm ecstatic.

"This goes all the way to the coast!" I say to Tabes as we walk up and down a little stretch of it.

We're wading through history. I picture the First Nations and European traders who once gutted it out on this trail with stinking loads of fish grease.

Tabes is more skeptical; he purses his lips after he's walked around a bit.

"You think so?" he says.

After some debate and closer map readings we lose confidence in this being the right trail. Time is running out; the machines aren't far away. I don't want to be responsible for severing this historic artery. The fate of a national landmark rests in the hands of two young men who can't read a map.

"Fuck," I say. "Where is it?"

Tabes has a pained look of indecision on his face.

"I dunno," he says. "Maybe we should back-track a bit."

We stumble back through the bush and now everything looks like the Alexander Mackenzie Grease

Trail—every abandoned deer track, every metre-long section free of brush. We make it to a meadow just up from the lake and a helicopter swings back and forth above us. At the top of the hill, the equipment is plowing through the meadow like it's scooping ice cream. The helicopter calls us on the radio.

"Go ahead," I say.

"How are you making out down there?"

"We're having trouble finding the trail."

"Okay, from up here it looks like it might be just a few feet uphill from you."

I see nothing nearby and I know they can't see anything from up there. Typical overhead riding around in choppers "seeing" everything, helping nobody.

We walk up to the machines. They've started their final descent toward the lake. We take out a bigger map and with the help of our action movie operator, we finally figure it out. The equipment group has to stop where it is. If we'd gone with our lakeshore instincts we would have sent the equipment right over the trail. To my embarrassment, overhead was right. Turns out you can see stuff from a helicopter.

In the afternoon, as we're making our way back up the hill, we meet up with the Castlegar Sentinels, a unit crew from the southeastern part of the province. A woman on the crew introduces herself and I get tongue-tied talking to her. She's gorgeous. We chat for a few minutes and then I say something like, "See you later down the line, then," as I walk away. My adolescent blabber is humiliating. An attractive woman is a shocking sight in the bush.

The fire continues to kick us around. In the late

afternoon, the beat-down is made official when overhead pulls us off the fire for the day. Dan gets on the radio to tell us to walk the guard back to the trucks. Along the way we're to collect fire gear, which is at risk of being burned up if left overnight. As we walk, the fire gets more active. We emerge from the timber onto a cutblock. Wind blows through the open landscape, whipping up dust from the dry earth and mixing it with the smoke. Somewhere in this grey soup, an orange spark is going to find its way across to the green side of the guard.

We reach a point where the smoke billowing across the guard is impenetrable. I'm about to veer left and avoid it by walking into the green side of the guard when I see some of our new people charge right into the grey curtain and disappear. Their move catches me off guard and I hurry into the smoke to try and fish them out. My eyes water and I time my breaths for when the smoke opens up a bit and I can see about a metre in front of me. When I get within shouting distance of the rookies I tell them to move into the cutblock on the green side of the guard. We get away from the fire and the smoke and gather in the cutblock.

"We don't have to walk through the smoke if it's like that," I say.

"I thought we had to collect the gear," one of them says.

"When it's that hairy we'll leave it. We don't have to kill ourselves."

The trudge continues, everyone now with runny eyes and noses. Some thunder cells pass nearby, the rumble like a distant building demolition. We're caught in a sliver of space between a lightning storm and a wildfire. It's rare weather and we're silent as we march to the trucks to finish up a day that started with promise but has ended in failure.

WE'RE LEAVING OUR Edenic rec site tonight to be assimilated into a fire camp set up along the main logging road that leads to Vanderhoof.

There's little narrative to draw on for the fire camp experience. They pop up in a variety of places and come in a number of sizes, housing anywhere from a few dozen to a few hundred firefighters. I write Sue a letter one night describing our current fire camp:

Dear Sue,

I'm writing you this from our filthy truck. A bunch of Ontario crews just showed up and they're all trying to park their rental trucks. Rob Zombie is playing on the radio, which makes me write faster. It's cold-ish and cloudy and starting to get dark. Across the road is the Cariboo Cougars hockey team bus, also for Ontario crews. After that is security. Security consists of hooded First Nations guys who ask your truck number once a day and generally look as though they've shouldered most of the brunt of 150-plus years of oppression. Security is rounded out by the kind of old guy who is aggressive in his story telling. He tells me about being in Vietnam and African peacekeeping and being a ranch hand in Australia. He tells me all this in the time it takes me to eat one pancake in the morning and for that reason I'm suspicious of his motives. Who would ever say that much about themselves before knowing the name of who they're saying it to? Anyway, once you're past security there are more trucks and then you are in the camp. Everything is semi-trailers—trailers with food, with computers, with toilets, with fire equipment. There is sewer wafting through it all and when it rains it gets muddy. Our sleeping arrangements are wall tents. The other day it rained. There were holes in the roof so I bagged up all my stuff and put it on my cot. Somebody said it was a real "veteran move." I've had this said to me a bunch of times and I'm not sure how I feel about it. Sad but good, better than nothing.

Anyway, just wanted to give you a rundown of what fire camp is like. It's kind of a bizarre place. It can feel a bit like maggots on a corpse, especially this one cuz it's in a gravel pit south of Vanderhoof. You get close to some sad people. I especially feel bad for the older ones, people in the same clothes as me, but they're moving toward sixty and they usually have some permanent injury—a bad limp or an eye patch.

AS A KID I was skeptical of those signs along the highway saying something as innocuous as a spark from a cigarette butt could start a fire. Today challenges my disbelief.

Our first task when we arrive at the fire's edge is to patrol the green side of a stretch of guard. It's the same stretch of land we were on during our smoky walk back to the trucks last night.

Less than five minutes into our patrol I see smoke

gathering nearby. I leave the patrol group and start planning how to work this section when it starts to boil over. It doesn't take long. The breeze picks up and I watch sparks the size of buttons sail over the guard. Within minutes there are several wisps of smoke popping up on the green side. Soon, the pencil-thin lines of smoke are chugging away like stacks on a container ship. The spot fires are in need of buckets. I spend the rest of the morning calling in helicopters to the dozens of fires that have sprouted across the guard. Because I'm walking in the wide-open space of a recent logging block, it's easy for the pilots to find my red shirt among the waist-high green trees.

A chopper approaches, its orange bucket descending like a pop fly I'm supposed to catch. The helicopter and bucket look tranquil at first but as they get closer, it gets loud and the wash from the rotors sends a hurricane-force blast of wind at the ground, throwing dust everywhere. I back off and let it make the drop. The 1,600 litres of water hitting the ground is like a detonation, but it's barely audible against the noise of the chopper. After each bucket drop the spot fires, now reduced to steam rising from mud, have the look of dormant geysers. There could still be heat in these spots, and I use my Pulaski to dig a hand guard around the craters. There are other hand tools used for firefighting—shovels and modified rakes. But the Pulaski is it; it is to firefighting as the hammer is to carpentry.

The tool was perfected by Ed Pulaski, a US Forest Service employee and firefighter. Pulaski famously saved the lives of most of his forty-five man crew when they became trapped in the Great Fire of 1910, a million-hectare blaze that covered parts of Montana, Idaho and

Washington state. He led his group to an abandoned mineshaft where they waited out the worst of the fire. Pulaski's story is one of few historical firefighting tales that is common knowledge among crews today.

Not far from where I'm digging, the patrol group has set up hose and is battling the fire in the timber.

I make a brief excursion to where they're working, arriving just as everything is falling apart. There's hose spread out all over the guard. Everybody is focused on trying to get water on fire, but it's too much to handle. They spray down a group of candling trees, unaware that another group of trees behind them is now engulfed in flames. The bush in this area is thick and it's hard to see where the simmering surface fire is going to make its next run into the treetops. Sometimes we hear these problem spots before we see them; fire tears through limb-loaded timber making a sound like a shirt being ripped in half. Like the nearby cutblock, there are spots here where the fire has crossed into the green side of the guard. Unlike the cutblock, the bush prevents us from identifying the smokes when they're small. Kara goes in after one of these spots but it's already putting out more smoke than a house fire. Their battle is lost, and I have to get back to check on what's happening with the escapes I've left behind.

Walking down the guard I hear Dan and Rob on the radio talking to each other. They're in agreement that the fire activity is getting worrisome.

From their conversation I gather that Blaine has been driving a truck back and forth down a short stretch of road for the last hour. He's making sure we won't be taken by surprise if the fire tries to block our access out.

While Dan and Rob talk I pick up my pace and head

toward Blaine. The truck comes up the road and I jump in. I want to see how the fire is looking on our access road. If it's anything like what I've seen so far today, we may need to get out of here sooner rather than later.

We drive down the access road. Flames are climbing up the trees and creeping through the grass beside the road. Smoke blocks the sun as patches of trees are consumed by fire. Out the window, mixed with the drone of our diesel engine, is the sucking noise of a fire demanding oxygen. Burnt sections of hose lay charred on the side of the road, melted into the dirt and curled up at the ends. In some spots we're surrounded on all sides by fire. My stomach hardens; this place is going to be a gale-force inferno in about ten minutes. In front of this inferno will be a dozen or so inexperienced Rangers.

Blaine reaches an area where the fire is ripping through trees close to the road. He slows down.

"Go fast," I say.

"Yeah?" he says.

"Yeah."

He drives through the intense heat but his alarm level is not nearly high enough. If we stay in these hot spots for too long the paint will peel off the truck. Our situation isn't desperate yet, but we're getting glimpses of what desperate will look like.

We get to the bottom of the road and I see Dan. He's standing at the door of his truck, radio in hand. He's looking at the fire now poking its head out of the tall timber, like the intermittent crash of waves onto a rocky shoreline. The fire is organizing itself to make a run at the guards we've been working since this morning. Blaine parks and I hop out.

"Are we good to go?" I ask Dan.

"Yeah."

Dan assigns a driver to each of the remaining trucks.

Blaine gets in the driver's seat of ours and I come around to the same side as him.

"Move," I say.

"What? Like move over?"

"Passenger side. I'm driving."

Blaine is fine and he probably could have gotten us out of there. But I'm prickly enough to strong-arm him out of the driver's seat. It's good to be in charge of your own life.

On the way out the fire activity continues to increase. Ten minutes ago the fire was just starting to tangle with the lowest branches; now it's climbing tree trunks as if they're fuses, reaching for the dynamite branches at the top. I'm going fifty kilometres an hour on a road that shouldn't be driven faster than thirty. It's a deactivated road, so there are huge dips into dry creek beds. Coming out of the dips, the truck spits rocks and fishtails. Waves of heat invade the cab. The windows are rolled up, but the sound of the fire penetrates the truck's interior and spurs me on.

We make it out of the fire, parking where I first met Blaine on the road. The remainder of the crew is now waiting there. I put the truck in park and everyone else jumps out, slamming the doors. I stay in my seat and let out a long sigh in the brief vacuum of silence that follows. We've lost again, a bigger loss than yesterday.

WITH THE TEAR the fire went on yesterday, the following morning starts slow. We languish in the hot trucks all morning. The blackflies outside are ferocious and we alternate between running the trucks to get some air conditioning and sitting in them until the heat becomes unbearable. We're parked in a clearing pushed out by bulldozers; the dry earth, once exposed, becomes the finest silt.

In Charlie truck, Tabes and I sit in the front seat; Kara sits in the back. Brian has gone off to chat elsewhere. We speculate and argue about everything from who the hottest people are in camp to who the best-dressed person on our crew is. Nothing is highbrow. This is the same for each truck; we all spill gossip like a supermarket checkout magazine. I suspect ours is the most malicious.

Eventually most of the crew gets called out to the fire line. Then it's just Brad, Tabes and me waiting to be pulled from the hot aquarium of our trucks. We'll be line locating, but not until management gets a read on what this fire did last night and what it might do today.

To pass the time we walk across the parking area; on the other side a few equipment operators stand near their machines. The operators have spent a lot of time in camps and are discussing camp food. First they talk about the best. This was in a place called Kemano, a camp servicing a remote dam close to the coast. The father of one of the operators worked there, and the operator spent a week there as young kid. "I thought I was in fuckin' Paris! Chefs with them floppy white hats."

The worst food story is from a camp south of Vanderhoof. "The cook was stealing bread from the Co-op dumpster," says one operator. He's not joking.

"You know what you do in that situation," says another operator. "You take it and throw it in the garbage."

He makes a strong throwing motion directed at the ground.

Talk turns to equipment. Beside us the circular blade of a feller buncher, an excavator-like machine used for cutting down trees, is rotating lazily in the wind. Its movement jogs the memory of one of the operators. Years ago his eight-year-old son lost a finger to an idle blade being pushed in the wind.

They also talk about how they managed to keep their drinking in check the night before.

"The beer was on the table but I didn't crack it," one says. "I'll need it later."

Brad loves talking to these guys. "There's two types of operator," Brad says as we walk back to the trucks. "Miserable and hackin' darts, or real happy. Either way, there are no regular operators."

Brad, whose dad drives truck for a living, says the older we get, the more like these operators we'll become.

"Those guys are so Canadian," Brad says. "They make the world go 'round."

His mood darkens when he compares our crew to them. "The Ministry is soft. Too many spoiled kids."

In the afternoon Tabes and I are shown a spot to line locate. It's odd when a fire sits down on a hot day like this. It usually means it's run into something it can't burn. We discover what this is when we step off the road. In the bush, dead trees lean at steep angles to the edge of a swampy area that has slowed the fire's growth.

We navigate this grey, inedible edge of the fire and it feels like we're fighting our way through an enormous compost bin. Clothing gets ripped, hard hats pulled off, laces come undone. The heat today makes it that much worse: it's nearly thirty degrees in this shady stretch of the forest. We hang ribbon for almost three hours before we turn around to walk back to the trucks.

We're sweating through our clothes by the time we extract ourselves from the mess and stumble onto the road.

It's getting dark as Tabes drives us back to camp. The setting sun mixed with the smoke and dust creates an eerie light. The landscape appears in hazy waves. We aren't moving along the surface of the Earth, we're suspended in it, driving nowhere, hypnotized by the light outside. The sweat cools on my shirt and I shiver for the first time in what feels like months.

THE NEXT MORNING Kara and I are line locating, ribboning our way down a steep meadow with a small marsh at the bottom. It's high overcast but still warm; a ridge of high pressure is breaking down. We walk across a cutblock and finish our line by linking it up to another guard built yesterday. Between the logging roads and the fire guards, the terrain is scratched up like an old CD. We walk back to the steep pitch and find that professional foresters have rerouted the guard away from the steepest parts of the hill.

This rerouting will take longer and it means the guard will be farther away from the fire. These foresters are following safe work standards, though. The equipment is only allowed to work up to a certain steepness of grade. I approach one of the machines; in the cab is Rob, the operator who compared Kemano to Paris.

The last bit of rookie week. Bronson is out front, Kelly right behind. A celebration and a crew-cooked meal to follow.

Smoke causes early darkness at Euchiniko Lake. The crew relaxes outside wall tents before bed.

Dinner on makeshift benches during the early days of Chelaslie. Putting some extra effort into making camp goes a long way for crew morale.

A couple of crew members watch a burn-off from the bush. Burning off can be risky, but there's no better way to shore up cat guards and prevent escape fires.

Dry conditions and big fires made burning off a common tactic in 2014. Kara is about to lay down another swath of fire from the guard.

Addison and Kara monitor a burn-off. Quads can be a lifesaver on long stretches of cat guard.

Smoke in the distance, not in our lungs. A rare clear ride on the Ootsa Lake barge.

For the cutters, it's a fine line between leaving things too big or too small for the swampers. Rob seems okay with the big stuff.

Tabes falls a spruce tree while a ground fire smoulders. He's at the end of a long escape route he cut for himself.

Chris is challenged to consume nothing but chocolate milk, twenty litres of it, for twenty-four hours. Pete tests the product beforehand. "Challenges" like this are common if there's ever a quiet day

Digging hand guard on a fuel free on the China Nose fire.
PETER SMIT PHOTO

Brad works through an area that needed extensive bucking. To avoid
a mess and to speed up hosing, fallen trees are often bucked.

Hanging out on the tailgate. Getting back to the trucks at the end of a long day in the bush feels glorious ...
GRAEME TABOR PHOTO

... as does the blast of wind from helicopter rotors, cooling you down or driving away hordes of mosquitos. Sometimes both.
GRAEME TABOR PHOTO

An Electra drops retardant on the Chelaslie River Fire.
PETER SMIT PHOTO

"Rank five plus." A flame front several kilometres long on the
Chelaslie River Fire. TROY WHITE PHOTO

"How steep of a grade will your company let you work?" I ask.

"Thirty-five percent."

"How steep could you work with no rules?"

Rob snaps his head back and grunts, "Fifty fer sure."

We watch the four machines build guard. The snap of tree trunks pierces the air. Engines the size of small apartments plow tons of earth, fall hundreds of trees and turn wilderness into dirt road at close to a walking pace. In the same way city dwellers gather to watch an excavator tear down an old building, we gather to watch machines claw fire guard through the forest. Sometimes we even cheer the process in a way we might if we were at a monster truck rally or a wrestling event. You could bring out the board of directors for Greenpeace and I'd wager they'd do the same.

The sky darkens in the late afternoon, so much so that if this were a townsite, the street lights would be coming on. This is partially due to smoke from another massive fire burning about fifty kilometres to the west. When we're walking out at the end of the day, Kara and I stop at the top of a hill and watch the fire across Euchiniko Lake. Spots flare up under low-hanging dark-blue smoke, and the candling trees emanate a light that looks like it could give you welder's flash. The whole scene is pretty Old Testament.

RAIN STARTS THIS morning as we drive to the fire. It hits our windshield, mixes with the dust, and is smeared away by the wipers. Outside, fat droplets bounce off the broad-leafed brush in the ditches. By the time we reach the staging area it's coming down hard. Hard enough for plans to change. We park the trucks and sink into our seats as the moist world presses down. I could stay in the warm truck all day as the rain cauterizes our fire, wrecking our good time.

But Dan shows up at our truck window and he's excited about the rain, planning to use this opportunity to stamp out the fire. It's not just work Dan's excited about; he also wants to talk about yesterday's "ops meeting." The ops meeting is a gathering for management each night after work. Along with upper management, crew supervisors and squad bosses have to attend, but our mental and physical exhaustion at the end of these days makes it tough, especially when warm dinner is being served in the next tent. Dan and I laugh at the agony of listening to the previous night's speeches from management.

"Did you see that guy?" Dan asks, referring to one particular manager who has the aura of some carpetbagger salesman vacationing at a Mexican resort.

"Oh my God," I say, starting to giggle at the thought.

"He's a lounge lizard!" says Dan and we both explode with laughter.

By the time Dan leaves, I'm back on board. The pissy front moving across the Interior is no longer oppressive. We can put out this fire.

Most of the Rangers spend the day cutting hose trail. Parts of the guard are way off the fire's edge so we'll cut into the bush, build a trail and move in close to the fire. The cutting groups work in steep ground, most of which is meadow. The sawyers are stooped down, running their saws back and forth, cutting grass. A lawn mower would be the better tool for many of the sections. The cutters' shirts are dark red from sweat; the dew makes the steep ground slippery and their pants are covered in a green film from the numerous times they've lost their footing and collapsed into the steep hillside.

Later in the day I run into Dan partway up the hill and he's anxious. He feels like we're the only people out here who actually want to put this fire out. This morning's rain has sent other crews into remission, cutting the pace in half. The fire may look like it's finished, but at fifteen thousand hectares and only a hundred or so people on the ground, there's bound to be a leak somewhere as soon as the sun comes out.

After seeing Dan I continue downhill, laying hose as I go. Looking south toward Euchiniko Lake, the same vantage point as yesterday, I have my first decent view of the fire since the rain started. There are no big columns of smoke; instead, a great mass of steam rises from the ground. The smell in the air is a more pungent version of the first drops of rain hitting a dusty road—wet fire, false security.

I pass Tom while laying hose. I haven't seen him a whole lot this season; the drive to the fit test has been our only lengthy visit. He's almost at the top of the hill where the fire starts hugging cat guard again. He knows he's near the end. The pendulum of his saw in the grass swings faster.

"It's not far at all," I say as I walk by.

Tom's face lights up.

"I live for that, man," he says of the push to finish something before the day is over.

After work, one of the Sentinel crew members catches me looking at Hot Sentinel in the dinner line. Fire camp is the perfect place for these kinds of Victorian era–style meaningful glances, and there's always some dazzling woman on another crew who is protected like a sister by her co-workers.

I say hi to Hot Sentinel as we walk by each other after dinner and we stop to chat. I'm nervous. The conversation moves at an unnatural pace—too fast. She talks about wanting to be deployed to California to fight fire in the redwoods and surf in the evening. In an effort to distract myself from the feminine figure in front of me, I picture what she's describing—her surfing and a mountain burning in the background. It's a bunch of hippie bullshit that would drive me mental in any other setting. But here in camp, any conversation with an attractive woman is a pleasant escape.

"IS THIS YOUR first time?" I shout at Nate, one of our rookies, as the helicopter blades cut through the air above us.

"No," he says. "When I was younger I flew in a helicopter."

Yesterday's rain has given us access to new sections of the fire that need to be walled in before the forest dries out again. This morning we're riding in helicopters to the remotest part of the fire. It's our first chopper ride of the year.

I don't believe Nate. His body language—the way he's looking around—betrays excitement. I remember my first ride in a helicopter with the unit crew. I tried to play it down as well, acting like I had a chopper in my garage. But when ten of us had strapped into the red Bell 212 flown by an ex–military pilot named Luka, I was so excited I had to bite my lip to stop smiling so much. The brazen flouting of gravity, the way you climb straight up, like the best trampoline bounce you've ever had.

As we lift off, Nate lets a little smile escape, but he quickly crushes it. His eyes dart around to make sure nobody saw. Then the smile comes back, uncontainable as we climb out of the trees.

The helicopter lands at the mossy edge of a small, circular pond, upsetting a raft of ducks and sending waterspouts shooting up from the pond's surface. It's always exciting to be dropped into remote land from a helicopter. Few if any people have walked this ground before us. We'll be pumping water from this pond and working our way west, cutting hose trail and spraying the fire's edge.

THE FOLLOWING DAY I'm working near Tom. His easy nature is a relief in a group often tightly wound. He possesses none of the jealousy and defensiveness that come with being on a testosterone-clogged firefighting crew. If he's passed up for so-called glory jobs—line locating or falling big timber—he shrugs it off, as happy building lunches as he would be building a helipad on a mountainside.

At the end of the day, we wait at the edge of the pond in tired silence for our flight back to the trucks. It's day ten and it's hard not to get sleepy as soon as there's a minute to sit down. Tom says, "Okay, I'm gonna try this out here." He's been working on a birthday rap for Brad, who turned twenty-five yesterday. He starts his rap and gets through a couple of verses before stumbling a bit. We're roaring with laughter and it pumps him up. He goes for a few more verses, incorporating memories from past fires with quotes from the Jim Carrey and Adam Sandler fodder of Brad's youth. I feel a lump in my throat. Tom has risked embarrassment and hasn't hidden behind sarcasm; it is all sincere. Just when working this fire was slipping into "one more bush job" status, he's made us feel part of something special.

THE EVENING OPS meetings are getting worse. They're filled with excessive thank you's and exclamations of what great work everyone is doing. When dealing with bureaucrats, there's an inverse relationship between the thank you's and the amount of work that's been done—the less work you do, the more you're thanked for it.

They're changing overhead teams for this fire. The new team is from back east, *way* back east. These men and women are here from the Maritimes and the differences between the BC ground crews and our new overlords are stark. The outgoing BC management team thanks them in advance (as if a well-compensated trip to the fairer half of the country were some kind of burden.) Dan, Warren, Rob and I are standing up in the back with wet boots and empty stomachs, making wide eyes at each other when the tangents get particularly obscure.

After the meeting I rush to get dinner and catch Kara and Chris finishing up their meal.

"I frickin' love you guys," I say as I sit down at the picnic table with my food.

"What do you mean?" asks Chris.

"It's just good to see you."

It is good to see them—something genuine after all the platitudes and lengthy self-justification speeches. Most people who become forest firefighters don't stick around longer than a few years and don't climb any higher than crew member status. These meetings are a good warning to stay away from what's up there. Best to do what Kara and Chris are likely to do: take advantage of the well-paid seasonal work for a few years, then move on.

WE CONTINUE FLYING into a remote part of the fire, building hose trail and hosing. We spend too much time in the grey area between hosing and patrolling. If you don't get every little hot spot when you're hosing, it's hard to cover any ground when patrolling. You end up digging at spots with your Pulaski for hours when water would have put everything out in a fraction of the time. On the other hand, hosing a quiet fire can become mostly an exercise in setting up and taking down your econo hose, moving along the fire's edge trying to find bits of heat. Moving econo from one thief to another along the hose lay often takes about half an hour.

I milk each spot more than I should, searching for pockets of white ash—the stuff that billows up in gritty clouds when you hit it with water, making it look like you've actually found heat. By noon the situation is bad; I'm dogging it hard. Not far away the fire catches a break and starts blowing up. Three days of drying and its rest period is over.

Out of boredom, I walk to the edge of the creek to get a better view of the fire. At the water's edge I see the sun shining on a grey wall of smoke rising from the timber. It's tough to be chained to this section of the fire when there's more exciting stuff happening elsewhere. The new overhead team won't change anything, though; they're in over their heads and quick to press the panic button.

"The panic button" is not a metaphor. It's the button for their mic in the helicopter. They spend a lot of their day monitoring the fire from the air. They have the ability to communicate with us on the ground through the radios strapped to our chest packs. From the air it's a disaster. Along the creek, it's quiet. We hose as if watering a vegetable garden.

The flight out tells a different story. There's fire everywhere. We're at the far eastern end of the burn. Looking west there's a nearly steady flame front. It slithers along the forest floor and weaves through the treetops, pulling unburned timber into the orange fray. There's nothing but fuel ahead, good fuel—open logging blocks littered with cured wood. We've missed our chance to get ahold of this beast after it was tranquilized by the rain. Why can't we kick these things hard enough when they're down?

ON OUR LAST day, we drive to main staging. Within its scraped parameters are at least fifty pickup trucks, several port-a-potties, heavy machinery, burnt-up fire gear and tens of millions of dollars' worth of helicopters.

We park and take in the scene. Wretched contractors smoke cigarettes and cough and spit. Others lounge on truck boxes or joke with their friends. Groups of managers scurry around with maps, trying to get people working on this fifteen-thousand-hectare-and-growing fire before the booze-free tailgate party gets too big a head of steam.

When these big groups of firefighters have hung around for too long, everything is on the table. There are physical contests of all sorts; often the games involve feats of strength, throwing rocks or jumping over something. A few years ago we were at a staging area set up in a provincial park outside Kelowna. Visiting crews from Ontario and Quebec made a ball out of duct tape and a bat out of a stick and played a game of baseball. Ontario versus Quebec, a friendly match complete with umpires.

A helicopter drops us off at the fire, and I help Dan and Brian carry hose to the end of a hose trail. We're moving away from the creek, closer to the flame front we saw from the helicopter last night. We're dragging previously used lengths. I try for eight at once, leaning in and straining everything. No movement. I remove two lengths and the shed weight lets me inch forward. The weight I'm dragging puts me horizontal to the hill. Sweat pours off my face. There's no target muscle group for this exercise; they're all involved or the train doesn't move. After making it up with one load, I head down for another. This time I carry rolled hose, four skewered on my Pulaski, four hanging from my shoulders. Some can carry six; others can carry as many as twelve. At about forty-five kilograms total, eight is good enough. I lift the Pulaski over my head and rest it across my shoulders. The hose couplings sink into my skin, grinding muscle into bone. I sweat worse on this second trip. With each step my legs quiver.

This is the best, cheapest way to fight fire—a person stringing hose through the bush. The tools we use to put out fire are almost as primitive as fire itself. The work has been resistant to change. Sure, we use water bombers and scan for heat from helicopters these days. But it's the hand tools and hose that still get the best results.

In other industry jobs the tools have changed. Everything is automated now. Until just after I was born, my family was in the logging business. The stories I heard as a kid were all about loggers. Loggers climbing up and down mountainsides, falling trees with chainsaws; mechanics working through the night underneath huge machines. It was all physical stuff. My parents and grandparents talked about the wiry strength of these workers, the amount of food they consumed. By the time I was born most of the work that required cafeteria trays to be used as dinner plates had dried up or been automated. Over time I've realized that's partly why I'm here. I missed that era, so hauling forty-five kilograms of hose up a hillside will have to suffice. Even if it's only on a handful of days a year.

At the top of the hill, the sweat is nearly launching itself out of my pores. It gathers at my eyebrows, the last line of defence before it gets to my eyes. There's a breach in the hairy dam and my eyes start to sting. If I move my arms to wipe my forehead, the precarious mess of hose on my back is ruined. Instead, I move forward, helpless to the sweat and doing nothing about the scores of mosquitoes sucking blood from my arms. I want Dan and Brian to see me carrying hose. It's vain but I want them to know I'm working.

After packing hose for most of the afternoon, I hardly have time to set up my econo before everyone is walking to the helipad for the final commute back to camp. When we get to the trucks the scene is mostly the same as when we arrived. The fire is still burning out of control and another ridge of high pressure has restored the heat. But the trucks are dustier and we are dirtier. Ash has gathered in the contours of my fingerprints. On my hands are a collection of cuts and scrapes coloured deep red and threatening infection. My blue pants are crusty from fourteen days of sweat and dirt. White streaks of dried salt run down parts of my red shirt, showing where the sweat collected.

Some weight is lifting. Pressure has built up over the last two weeks. A pressure to perform—to do well by Dan and the crew. Though it's my fourth season as a squad boss, there's no complacency. The more I start to recognize myself as some sort of role model in this unit, the more concerned I am with maintaining that status. It makes for exhausting deployments. I relax a little, but not much. What if we get called again? Or worse, what if we don't?

Anticipation builds on the drive back to civilization. Tabes makes more jokes than usual. Any passing thought is spoken. Even Brian the rookie is being treated well. The satellite radio is tuned to Jimmy Buffett's Radio Margaritaville and we ride a wave of tropical music down the rolling hills to Vanderhoof.

When we get there I drop $25.55 on a tin of chew at 7-Eleven, then another $10.18 at the Co-op on fruits and vegetables. It's a contradiction in health and I don't need any of the items, but that's not the goal of this exercise—the goal is to spend money.

Spending money makes you feel powerful. It quickly converts endless hours slogging through the bush into joy. Loggers and other industry workers used to stay in camps for long periods of time. They'd make as much money in a few months as many people make in a year. Then they'd go to town and spend it all, sometimes so fast they'd have to go back to work early. The highs and lows of that life have shrunk in the twenty-first century, but they're still there. I feel a fraction of that impulsive behaviour this morning in Vanderhoof.

When we get back to the base we do some organizing before we go home. The excitement builds as the crew gets closer to a few days of better sleep, fresh food, alcohol and sex.

After a brief meeting Dan dismisses the crew except for the squad bosses. Everyone races out the door, out of the base parking lot, on to freedom.

Rob, Warren and I finish up some paperwork. Rob and Warren finish fast. They're both newly in love and anxious to leave.

It's down to Dan and me. I finish my work and walk across the lawn to my locker.

It's a cloudy evening and the wind meanders through the pine flats; it makes a low hum as it bounces off the tin siding of the buildings before continuing out over the sparse forest. Emptiness rushes in—no girlfriend here, no desire to party, no need for alcohol. I grab my backpack and wander back over to the crew room. The room is so quiet I can hear the buzz of the overhead lights. Dan is hunched at the computer. He swivels in his chair and rolls back when I enter.

"That was a good tour, Willy. I was leaning on you pretty hard there."

"Thanks, man. Yeah, felt good. Felt like we were on the same page."

I thought he might say something like this. It still feels good that he said it.

Dan waves at me from his truck as he leaves. I follow him, driving slow on the straight stretch of gravel road leading away from the base.

4
Chelaslie River Fire Tour #1

July 31–August 13, 2014

Back in May of 2006 the Telkwa Rangers were called to a fire south of Vanderhoof. The snow had been off the ground for only a couple of weeks. But a short warm spell meant everything dead from the previous fall was cured and ready to burn.

On our way to this fire we were held up by a section of road that had mysteriously become muddy. Each time a truck drove through the section the road sank, like a rotten roof under a heavy snow load. By the time the last trucks drove through, the road was a grey slurry only passable in four-wheel drive with the pedal to the floor.

At first we thought it was just spring melt. But after a while we realized what we were seeing was a consequence of the mountain pine beetle infestation. For millenniums these beetles kept forests in check, killing off older trees and leaving their dry husks to nurture the soil or burn up in rejuvenating wildfires. But in the midnineties an infestation of the beetles took off and grew so big that it eventually killed off forests covering a fifth of the province's land mass.

One reason beetle populations got so out of hand was climate change; they thrive in warmer temperatures. Another was overambitious forest firefighting. For the better part of a century, the policy was to fight any forest fire, no matter how remote. Thus, the number of mature pines susceptible to beetles had never been greater.

I remember standing next to that bit of road trying to wrap my head around how this plague had caused the entire water table to rise. How the road had turned to porridge because there weren't enough living trees to sop up spring snowmelt.

After a few years, beetle-killed trees shed their needles. "Grey and dead" is what we call these ghost pines. There are billions of them in the Interior of BC. Desert-dry standing sticks of firewood.

This was the state of the forest in the Ootsa Lake area when, on July 8, 2014, at 6:23 p.m., Nancy Dogleon, one of only a few people still employed as fire lookouts in BC, made a call to the Northwest Fire Centre. Lightning had started a fire in a remote area. For several days the Ministry left it alone, part of an effort in recent years to let wildfires burn free when possible.

DESPITE ITS ATTEMPT to leave it alone, three weeks after Nancy's call, the Ministry sends us to what is now known as the Chelaslie River fire.

We drive south from the town of Burns Lake, crossing François Lake on the government-run ferry. Another two hours of driving and we've reached a second water crossing. This one will take us to the south side of Ootsa Lake and the western flank of the fire.

We drive onto a cable barge that used to serve the logging industry but doesn't see much action now. Isaac, a skinny man with short, thinning hair, runs the barge. He revs up a tugboat that's tethered to the side of the barge and then leaves it alone, letting the tugboat push the barge, which is attached to a cable spanning the lake. Once we're moving, Isaac walks the deck of the barge with a clipboard, introducing himself to people at random and writing down their names like he's the sole customs agent for all of this desolate land.

On the far side of the lake, a stand of dead trees sticks up in the water. The trees are a reminder that Ootsa Lake is a man-made reservoir, part of an ambitious project to supply power to an aluminum smelter three hundred kilometres to the west. In the 1940s, engineers built a series of dams and reversed the direction of an entire drainage system that at one time flowed into the Fraser River. In the making of the reservoir, the small Cheslatta First Nation was uprooted. Their village and traditional territory were drowned only a week after people were told to leave. They requested that the graves in the village be relocated, but it didn't happen. When the water came, the coffins of their ancestors bobbed to the surface.

This whole commute has had the aura of a medieval quest. The feeling reaches its peak when the barge ramp drops like a drawbridge and we drive off into a remote wilderness, seeking out the edge of a massive inferno.

We try to get our bearings on where the fire ends, but every time we gain a new vantage point there's a new horizon of smoky edge. The only clear message we get is how undermanned this fire is. Its fifty-thousand-hectare size (and growing) is split into two halves. To the east several unit crews are working from an established fire camp. On the western half, we've joined a few contract firefighters and two managers from the Ministry. The managers have been sent from the east camp seemingly as an offering—as if the presence of their red shirts alone might encourage the fire to behave. We have two dozen people with which to monitor 250 square kilometres of fire. Our presence here will have as much impact as a single car on global warming.

We camp at an abandoned log-sorting site at the edge of Ootsa Lake. It's a series of gravel terraces framed by alder trees. Terraced construction is often associated with fertility and the remnants of great civilizations. Not here. These terraces look more like the remnants of some long-ago apocalypse—they're barren save for the bits of twisted steel and cables poking out from the gravel. Still, the spot is relatively flat, open and next to water. Tents are set up, firewood gathered, benches built and within an hour we have a camping spot.

AT FIVE IN the morning I wake up and poke my head out of the tent. Ootsa Lake is calm and dark purple in the pre-dawn light. The trees in the lake look like something imprisoned. It's cold and I decide to stay in my sleeping bag, drifting in and out of sleep until a flurry of nearby alarm clocks goes off at six. In the confines of the tent I put on as many clothes as are available before wriggling out the door with the smoothness of a birth. Once outside I hurry to put on more clothes. I eat cold breakfast cereal with the bowl in my lap, alternately warming each hand over boiling dishwater. There's no ceremony in this breakfast, no coffee, no oatmeal for warmth, no fire inviting a few minutes of staring before we get ready.

Our best option for beating the cold is to get to work. We set up hose on a long cat guard built before we arrived. In the distance the fire is threshing huge tracts of forest with algorithmic efficiency. The dry conditions and the fire's aggressive behaviour at this early hour are a good indicator that this workday will be cut short. We stand around on the guard, chewing and baking in the hot sun. Later, we're pulled back to the trucks, as there's nothing useful we can do.

My squad cooks dinner in the evening, and this will be the most anyone on the Rangers achieves today.

When we get to camp, Tabes and I wrestle out an old grader blade partially buried in the gravel. It's rusty and caked with dirt. It will be our barbecue. We scrub it off with steel wool, clean it with cola and place it over the fire. Once we've prepped the rest of the dinner and the crew has returned from the fire line, we put steaks on our makeshift grill, eight at a time.

Dan holds the after-work meeting while we grill the steaks. The crew is paying more attention to us than to Dan's end-of-day spiel. The pressure of getting the meat right is immense. Is it too rare? Or the ultimate shame— is it too cooked? It turns out fine, and Tabes and I share a moment when it's all done. A stern nod to each other acknowledges that our integrity as men has never been stronger.

We sit around the fire in the gathering smoky dark, enjoying laptop-sized servings of red meat. Yesterday we were so clean we looked like actors cast as firefighters. But after one half-assed day we look similar to the iron grill we dredged from the log dump.

THE FIRE ACTIVITY we saw from a distance yesterday ended up challenging the stretch of land we were hosing. The next morning we discover that an escape has burned into some old slash piles.

Partway through the day, help arrives in the form of an old excavator jostling down the guard. In the cab is a man who's well into his sixties and tall. He's sitting down but still a presence. When I reach up to shake his hand, he grabs it like a bear swiping at a fish. It envelopes mine and he gives his version of a gentle squeeze. I press back but my bony fingers barely register against his hands. He introduces himself as Carl. I ask him what he does when he's not working fires.

"Well, I'm tired, not *re*-tired," he says. He tells me he owns a ranch just off the François Lake ferry dock, where he lives with "the wife."

"You go past the white church and it's just two driveways up." I don't know what he's talking about but appreciate being spoken to like a local.

Carl owns a bunch of equipment and during fire season he contracts himself and his machinery out to the Ministry. His jeans are worn thin, as is his denim shirt. His chest hair looks like it could plane wood.

I want to know more about his life, as it seems tidy and God-fearing.

"So is it just you and the wife at the ranch?" I ask. I've already built a vision of it—eggs for breakfast, farm work all day, maybe he watches the news with the wife in the evening.

"Yeah, it is," he says in a satisfied tone.

The sample size is small, but people living on the south side of François Lake seem to be from a different time altogether. A time of fire lookouts, primitive machinery, a rifle behind the front seat.

We return to camp in the evening to find our tents coated in ash. There's a patch of blue sky directly above the lake, but all around us in the middle distance, smoke rises up from the forest. Its movement looks like the slow billowing of a theatre curtain.

We're completely alone here and it's firefighting at its best. We're free. No overhead, no eyes in the sky, no rules. It's us and a massive fire.

AS AN EXPRESSION of this freedom, Tabes, Kara and I spend all of the next day burning off.

Burning off is rarely done in Canada. BC is the only place it happens with regularity. But even in BC it's viewed with trepidation. If it goes wrong, there could be major consequences. The X factor is wind; a change in wind direction, however small, can be a huge problem, sending burning shrapnel to the green side of the guard, propelling the fire from marginally contained status to starting again from scratch. Combine that with a government culture of caution and cover-your-ass, and burning off is pretty rare. Since I joined the Rangers we've done it on a large scale about half a dozen times.

Our burn-off doesn't work at first. We're trying to light second-growth trees, but the young trees retain too much water to be as potently flammable as the rest of the forest. Our first attempt is in a stand of spruce, bushy and stout. Attempting to burn it is like attempting to burn a wet dishcloth.

Dan calls on the radio. "Ranger Charlie, it's Ranger One."

"Go ahead."

"How are things going on your end there, Willy?"

"Not great. We're just about to try a patch of pine that's a bit more mature."

"Okay, head back over here if you're not having success."

Dan said earlier we should help the rest of the crew cut hose trail if we can't get anything to light. But burning off is way more exciting than cutting hose trail and I badly want this to work. Our squad has spent the last three days on the quietest part of the fire. Meanwhile, the rest of the crew has been working an area that's challenging the guard. They're watching water bombers at work, driving ATVs and cutting down trees. Tabes, Kara and I finally have a chance to be the ones who are envied.

We walk into the forest with our drip torches—metal canisters with a spout at the top where gasoline comes out in a controlled drip. In the distance, there's a thrumming of chainsaws and pumps. The pine trees are much bigger than the spruce, well on their way to being valuable timber again. Another thirty years and the forest companies will see a return on their investment. The orderly rows of trees, combined with knowledge of their value, intensifies the mood. Leaving a thin strip of trees burnt to death without even slowing the fire will look stupid.

As soon as we tilt our drip torches to the ground, fire expands along the forest floor in an almost liquid way. Clumps of needle-heavy branches catch fire, creating a pleasing sound like breaking glass. Sap hisses and bubbles on the stems. Five minutes later we have to shout through the roar of the fire. Communication matters here and for once it's all serious.

The burn is spreading out of sight. Urged on by the wind, it rushes deep into the cutblock. Up the hill from where we stand, a smouldering ground fire, inching downhill, will meet our burn-off. The whole plantation of pine trees will flare up and die down with the ferocity and harmlessness of a distant dying star.

ELSEWHERE, THE FIRE is challenging the few attempts at containment we've made. The hose trail the other group has been working on should be fine. But farther away the fire is crossing guards like they're speed bumps.

We're told the fire activity is too intense and are redirected back to camp so we can pack up the gear. A proper fire camp is being constructed on the north side of Ootsa Lake a few kilometres from the barge landing.

Dan and a few others are away from the crew and it's up to the rest of us to break down our campsite at the log dump. Dan has instructed me to pack up his tent. Between that and overseeing the breakdown of camp, it's a busy afternoon.

We arrive at the new camp late in the evening and set up in wall tents in the dark. The camp is built on the remnants of an old forestry camp. A few buildings are still standing, painted brown with simple white trim and signage like "Fire equipment" and "Fuel shed." They look like pieces from a Lego set. The main buildings have been bulldozed and they sit in a rotting heap in the middle of the clearing. The clearing itself is a pit dug down to bedrock and surrounded by tall trees. On one side is a steep drop into the lake; on the other a narrow band of trees separates it from the road.

Near the centre of the camp is a cluster of wall tents and semi-trailers. Orbiting the centre and scattered among the trees are individual tents belonging to contractors such as fallers and first-aiders.

There are small trees throughout camp—saplings growing out of the rocky earth. While we're here, this little ecosystem trying to reclaim its lost ground is trampled by a thousand late-night pee breaks and an ever-increasing amount of white splatters of toothpaste.

The food here is slightly better than at the last camp. There are fewer tuna sandwiches and less of a lineup to get food. The eggs appear to have been in actual shells that morning. The oatmeal is better and they've added a few extra juice crystals to each vat of juice, upping the flavour by three parts per million.

NOBODY WANTS TO be in Dan's tent. Time alone is unattainable, so time away from the boss is some consolation. I end up being one of ten people in Dan's tent. After dinner a few of us are lying on our cots chatting. Before I drift off to sleep, Dan comes through the door.

"Has anyone seen those dumbbells?"

He's brought some dumbbells along for exercising in case of downtime.

"I hid them on you. I didn't want your arms getting any bigger," I say.

He counters by saying, "The trucks are fuckin' disgusting."

The tent goes dead quiet and the night ends.

In the morning Dan addresses the crew about the messy trucks. I can't stay silent. Work has been like canoeing through rapids these last few days—just hang on to the gear and survive. When he's finished his speech I add that yesterday was an unusual day. We packed up the old camp in a rush and got to the new one at a late hour. I try to calm my voice when I'm talking, but it still comes out quivering.

"Having things clean is always the goal," I say, "and we've been good all year, but we should always shoot for clean trucks."

I shouldn't have added the *but*, and Dan pounces on it.

"Exactly, so there's no excuse," he says.

I have a good head of steam for most of the morning. We're burning off again and I don't rush it. Tabes, Kara and I visit at the tailgate before going to work. Gradually I calm down but as I do, burning off starts to lose its lustre. We're hot, tangled in the bushes, and the gas fumes feel like they've singed off all my nose hair. The heat draws the strength right out of you. Sometimes, after emptying your torch on a particularly hot section, you need water almost as badly as you need air. By the afternoon my nose and throat are a burned-out cavern. All day long, my nose drips snot that's a diseased yellow colour.

At the end of our burn-off, close to where the guard turns into hose trail, there are a series of slash piles strewn among the pine trees. The piles consist of greying logs unfit for the sawmill. We light them up and the inferno is enormous.

While the slash piles burn, I run to the trucks and grab the garbage we've been carting around for the last four days. With garbage bag in hand I wind up like I'm doing the hammer throw and release the trash into the fire. The top of the bag is open and the last thing I see before it disappears is a plastic muffin tray. It rises out of the top of the bag, propped up in the air by heat. Then, while still on an upward trajectory, it crinkles up and is vaporized.

With our burn-off complete we join the hose trail group. I see Blaine on the trail, and we talk about the morning's scolding. I'm angry at Dan all over again.

I go back to my hose but I don't open the nozzle. I eat a granola bar alone, then take a walk in the woods, revelling in the feeble act of rebellion.

We face a lot of piss-offs in this job—similar to any job, I suppose. It's the way things go. People's pride is always clipped down until it's almost non-existent.

THE SMOKE IS dense and it's warm in the forest. I do a couple of hikes along our hose trail and feel it in my lungs and head. These days the sun occasionally makes an appearance around noon. But it's gone within an hour, becoming a dim bulb behind miles of smoke, like a flashlight going dead.

The smoke looks like a dense fog but feels nothing like it. It's stifling and dry, making noses runny, causing headaches, coughs, fatigue. As a kid, I went through a dinosaur-loving phase. These days remind me of how the books I read portrayed Earth after the dino-killing meteor had struck—an orange-grey haze devoid of vibrancy.

We've been out here for nearly a week and there's no cell service, no weather report. Just smoke sitting on top of us, coating our lungs, dulling our senses, compounding our isolation.

We've been in more isolated spots but somehow this area feels more remote. Perhaps it's because down here, in the middle of the province, we're surrounded on all sides by civilization. It's unreachable, but it's there. This makes you feel farther away than if you're working on fires in the north of the province, where there's nothing in almost every direction. Entire time zones of emptiness.

This morning a big group has been assigned to work on a hose trail. The trail will stretch deep into old-growth forest, away from logging activity. Away, possibly, from any human footprints ever. Some people have been working on this hose trail for five days straight. Right now the sheer physicality of our work rivals any tough job you can name.

I'm supposed to be hosing but I walk to the end of the trail to get a sense of where we're going. I also want to take stock of our energy reserves.

After two kilometres of walking, I come upon the cutting and swamping groups getting ready for the day. They're slow; they could have been working by now, but there's no sheepishness or defiance from them—they're just tired.

Tom is the most experienced of the group; in his second year he did five deployments in a row. I ask him how he's feeling.

"Honestly man, tired. But I'm gonna be a freight train when I get going."

My next mission is to check the water source for the hose trail we're constructing. We've found a small lake that sits above where we're working. I walk the trail up to the lake; the climb would be easy if it weren't for the heat and smoke. The air is thick as cement. Any movement feels sticky and anything beyond a normal heart rate can't be sustained for long.

The lake is a dab of fresh water on an otherwise blackened, parched landscape. Trees line the water's edge and organic sludge wallows inches below the surface. Given a hot enough day (today might work), it feels like the whole thing could just evaporate.

We have two pumps coming out of the lake. One of the pumps has died; the other shudders along, drawing a thread of water through blackened hose down to the crew below. I start the dead pump and watch it work for a while before leaving the site.

On the way down I see Tabes. At thirty metres away his figure is distorted by the waves of smoke and heat coming off the black earth. His head is down and he's walking through the ash with all his saw gear. When he gets closer, he lifts his head; enough sweat runs off his face to brine fish. We stand next to each other for a bit before saying anything.

"Hey," I say.

"Hey," he says back.

I tell him about the pump trouble; he's heard and is on his way to fix it. I grab some of his gear and we walk up to the pump site. The formerly dead pump is still running and we spend a few minutes hanging around, not wanting to leave the lake only to have the pumps die, necessitating another long walk. It's a fine line between laziness and prudence on these long hose lays. This one stretches nearly three kilometres, about ninety lengths.

I leave Tabes and walk back down the trail. The fire is coming alive; stretches that were hosed down yesterday have made miraculous recoveries. Our hose trail could not be more vulnerable. In conditions like this, fire takes hold deep underground. It's early August and the forests have been drying for a few months. The summer heat first draws moisture out of everything on the ground and in the trees. Then it starts vacuuming moisture out of the soil, making things like tree roots susceptible to fire. Right now you could spend hours hosing a spot the size of a queen-sized mattress and still not have put it out.

Ahead of me on the hose trail, I see a thick cauldron of smoke. Some fire has found good fuel and is making a move across our trail. The heat has punctured the hose and a jet of water pulverizes the ground, exposing roots and boulders. I grab the hose on each side of the puncture and spray as much of the growing fire as possible. Steam fills the air. Pete, a second-year who's the younger brother of Blaine, arrives through the hot fog with some hose. Where Blaine is reserved and cautious, Pete is boisterous and impulsive, earning him the nickname Pistols Pete. We change out the leak and continue down the trail.

It's not even noon and radio traffic is on the rise. There are helicopters bucketing and water bombers on the way. We hear air traffic roar by, and the sound of gathering fire is briefly replaced with the sound of pistons and turbines. Fire owns every sense. Everything we see, hear, smell, feel—it's all wildfire.

Pete and I walk past the end of the hose trail and see a barren cutblock ahead. The smoke is starting to clear and the wind is picking up as we reach the hottest part of the day. Somewhere around this block we're supposed to connect up with another guard. The two fire guards linked by our hose trail in the middle will be nearly ten kilometres long.

The radio chatter increases. There's still only one helicopter with one manager attached to this side of the fire and he's talking non-stop, commenting on the fire as if it were a horse race. We can't see much more from the cutblock but we linger for a bit, hoping to catch a glimpse of some action—water bombers or a growing column of smoke.

I open an energy drink that's been in my bag for a few days; the writing on the can is barely visible. It's a Beaver Buzz, Saskatoon Berry Flavour. In faded excitement, the can proclaims it's "Dam Good!"

Energy drink culture is an embarrassment. But it's the perfect food for firefighting. Sleep deprivation can become severe on these fires—we've worked almost ninety hours in the past six days, and it's starting to show. There's no time to sip coffee, nor is there any access to it, not that we'd want it in this heat. Energy drinks give us something cool (or at least not hot) and caffeinated.

"You want some?" I say to Pete.

"Uhhh, you know what? Yeah, man, I do."

"Of course you do!" For some reason, I feel it's necessary to convince Pete of his own fatigue.

Dan calls us back to the trucks over the radio. As we start walking, an Electra air tanker comes over the treetops, screaming into our lives, its ancient propellers working against the dead weight of the red fire retardant sitting in its belly.

Electras were built by Lockheed. A limited number were produced in the 1950s and very few are still in service today. In the early 1960s, one of the Electras now used by the Ministry was the team plane for the Los

Angeles Dodgers. Given how it looks and sounds today, it's hard to imagine it as a luxury aircraft.

"Fu-ckin' right," I say as we watch the plane from the cutblock.

"That thing is massive," he says.

Once the Electra's fire retardant is dropped on the forest it can slow (*not* stop) a fire for several hours. This makes it especially useful if you're trying to hinder a fire that could be a bigger problem down the road.

In addition to the Electras, the Ministry uses a fleet of single-prop Fire Boss planes on floats that can drop both retardant and water on a fire. If there's a lake near the fire (common in BC), the Fire Boss can make several water drops in the space of an hour. By contrast the Electra has to go all the way back to one of the seventeen air tanker bases in the province to fill up with retardant.

We catch up with the rest of the group on the hose trail and head for the trucks. Some jokingly cuss about the ground we'll probably lose after we leave. Nobody cares too much, though—losing the ground was inevitable.

From our parking spot we see two mountains of smoke puncturing the blue sky like the fallout of an atomic bomb. It's rank six fire, but it's an unwritten rule in the Ministry that you never call anything rank six. Managers seeing rank six fire from the air use terms like "rank five plus" to fill in the gap. A dispatcher at the fire centre named Rory later tells me how people were reacting to this fire from the management side.

"You'd have air attack officers, guys who had seen crazy fire for twenty-five to thirty years…blown away by the fire activity. They were calling in saying they were seeing 'rank five plus, plus, plus.'"

When we get on the barge to go back to camp, we see smoke hanging in the air, looking too dense to be up there, as if invisible wires are suspending it in the sky.

It gets dark early. At camp, people's thoughts are preoccupied with what havoc the fire is wreaking on the other side of Ootsa Lake. It feels like anything is possible, like if this keeps up not even a kilometre-wide body of water separating us from the fire will be enough. The ferocity of the fire makes it feel like we're not actually done work for the day, just hiding out and regrouping. There's an animal edginess that pervades camp.

I'M ON THE first barge over the next morning with the machine operators, who are nervous about their equipment, which was abandoned in haste yesterday afternoon.

As we cross, the operators scan the southern shore of the lake looking for some indication their machines are all right. The initial signs look good. The timber is still green. Beyond that, smoke hangs murky above the trees.

Everything is fine. At its closest, fire found the edge of safety zones where equipment was parked.

The fire didn't run too far over our hose trail, but far enough to force us out and start work on a new plan. We drag what's left of the hose off the trail, roll it up and move it to a new spot a few kilometres to the west. Kara drives us there. We're eating chips. She requests a water and Tabes, a stranger to the back seat, cracks one open and passes it to her. This is the first time Kara's driven the truck this year. Eventually she cracks.

"This is the best day ever!" she says as she reaches back for a handful of chips.

Once the hose is moved we start in on a new plan,

a tactic we've heard of but never used or seen before. We're calling it Superguard. Bulldozers and excavators will push in a regular guard. Then the operators will turn their machines ninety degrees and push the guard out another thirty metres or so. The result is a cat guard about five times wider than normal. Its dimensions are similar to those of a soccer pitch, only it goes on for kilometres.

Superguard makes us feel small, a theme on this fire. Water is the one tool we use that gives us a sense of power. Hosing from a guard or a hose trail is where ground crews like us accomplish the most relative to all the machinery working the air and the ground. Superguard has erased that sense of power. If anything's going to stop this fire, it's going to be the obscene width of this guard, not our piddly garden hose.

Once we've laid hose out along the Superguard, I go start the pumps. Time to test what we've assembled. We're pumping from another lake, one that has a wide grassy meadow leading to its edge. It's bigger and bluer, more inviting than the fire-choked sludge pond we were pumping from yesterday. A breeze moves through the parched yellow meadow grass and the air coming off the water is clean and crisp. I sit down on the bank, the first time my mind and body have been shut down all day. In less than a minute my eyes start to close. I stand up and pull the pump's starter cord and it starts easily.

Helicopters are bucketing from the lake. An orange bucket hits the water with a crash and sinks in a geyser of bubbles. It rests underwater for a second before the line is pulled taut and the chopper shudders with the effort of lifting it into the air. The helicopters, once flashy with frequent washing, now show their own signs of fatigue, their tail booms dark with the residue of exhaust.

According to a BC Forest Service newsletter, helicopters were first used for firefighting in BC in 1956. That was when an early model Bell helicopter dropped a 225-kilogram payload of fire retardant on a small fire started by a lightning strike. There was no bucket for these drops; instead something called a "#10 size bag" was filled with a gallon of fire retardant. These bags were launched by hand from the back seat of the machine. On aiming the bags, the newsletter says it was "imperative to…keep one's eye on the target." It further explains difficulties caused by "violent updrafts which occur on the mountainsides during a hot summer afternoon."

One thing hasn't changed—the worry about the cost. But, like seeing ten-cent coffee on an old diner menu, the cost back then seems quaint—$100 an hour to operate the machine. These days a mid-size helicopter hires out for around $4,000 an hour.

As I'm walking up to the second pump in the line, a woman from another crew heads down to the meadow. A unit crew from Williams Lake is pumping water out of this lake as well. The loud pumps negate any need for a greeting between the two of us. She has good posture and I can see the curve of her red lips. I haven't thought about curves or lips or women for a long time. Here is one. Walking through this big empty meadow, staring straight ahead. For a few delusional seconds my thoughts turn to the most improbable scenarios involving me and the Woman in the Meadow. Then consciousness rushes in and I remember a lot of things, like the fact that I'm in a relationship. I swallow my lust and wave. Her return wave is decisively cold.

As I walk back across the meadow, an Electra bomber appears low over the treetops on the other side of the lake. There's talk on the radio. Dan is near the water bombers and has to move his group out of the area. I realize there's a ton of smoke coming out of the forest near our old hose trail, and it's leaning toward Dan's group.

The Electra makes a few passes, each one slower and lower. It looks tired in the air. The bird-dog plane that guides it seems to be dragging it by an invisible leash while the big plane moans along behind. The two planes circle around and fly lower again. On the final pass the Electra is so impossibly close to the ground that it disappears into the trees. It's hard to believe it hasn't crashed, and even the noise stops for a second and is replaced by the steady piston throbbing of our water pumps. I stare at the horizon and don't breathe. After what feels like several minutes, the Electra resurfaces, arcing into the blue sky with a thundering roar and the last of the red retardant misting off its tail. The mechanical glory of all this is as stimulating as anyone's good posture or red lips.

The fire doesn't make any aggressive moves toward us and the day ends uneventfully. There's talk of burning off from the Superguard. There's also talk of cutting a hose trail off the Superguard. We've never been involved in a fire so decidedly infinite. We seem to have been dropped into the middle of something eternal, a battle that was being fought long before we arrived and that will continue long after. We don't think about how to stop this fire, but rather how to set things up for somebody else to stop this fire.

In the evening we shovel food in as fast as we can. It hits my mouth and it feels like my stomach is going to reach up and grab it before it's been chewed.

I finish dinner and do several laps of the cereal shelf. First a bowl of Froot Loops, then several more bowls of Froot Loops. Partway through my second bowl, somebody gets up from the table and I move out of the way and spill half the bowl all over myself. Milk and cereal cover my lap. I do nothing about it, and nobody makes a fuss. Out here it doesn't matter that my pants are soaked in milk and it doesn't matter that Froot Loops are ground into my pants. Like the fatigue shown in the exhaust and retardant stains on the aircraft working this fire, I too am tired and don't care about comfort or appearances.

THE NEXT AFTERNOON there's chaos on the Superguard— confusion with tasks, bad tasks, tired crew. We start by hosing down parts of the fire burning close to the guard. Then we're told there's a helicopter burn-off planned for this section and we should hose down the green side. Hosing the green side would make a strong showing on a list of useless firefighting tasks. We're supposed to direct our water to the green side of the guard in an attempt to prevent errant burn-off sparks from taking hold there. "Just raise the RH [relative humidity] in the area" is one way I've heard it described by a higher-up. But it's said in a winking tone I can't return because there's nothing winky about being stuck watering healthy forest.

Dan is away the whole day and when he comes by in the afternoon he rips a leaky pump out of the line even though Tabes tells him he had the water delivery system as well tuned as possible given the shoddy equipment.

Next he's upset at Tom for not following instructions given earlier in the day. His actions catch me off guard and my tiredness allows anger to seep in with little resistance.

Once Dan's gone I try to go back to hosing the green, but I'm too rattled. I need to vent to someone. The person I find is Ian, a YEP in 2012 and a rookie last year. Ian is a tall, soft-spoken farm kid with red-blond hair. Easy to picture in overalls with a piece of hay in his teeth. But there's a quiet intelligence about him. He conveys real concern for my blazing rant. When I'm at the height of my monologue he stops me and says, "Well, I guess we'll just have to do what we can to get through it." I wish I'd been as reasonable when I was twenty.

OVERNIGHT IT RAINS lightly and the moisture makes us bold enough to cut a hose trail off the Superguard to access the fire's edge. The cutting groups start in the morning while I walk ahead to see what kind of ground we'll cover. I end up spending the rest of the day hanging ribbons to guide the cutters. It's flat and featureless in this part of the woods.

I also make the mistake of leaving my backpack at the beginning of our hose trail. Given the work and the fatigue, food and water are needed at least every hour. I'm gone for almost three and feel awful and weak by the time I get back to the guard to pick up my bag. I eat leftover chicken so fast that it gets jammed up in my throat. I mix the next bite with water, changing the consistency from meat dust to meat mud.

I finish the snack and go back to the trail, trying to figure out where the burn goes. Imagining the edge of a forest fire, you'd think there would be an expanse of scorched earth that makes a clean, dead stop where the fire dies down. This isn't the case. Fire twists and turns along the forest floor in a way that can be maddening to follow. It's like seeing a coastline on Google Maps: zoomed out it looks relatively smooth, but zoom in and it becomes impossibly jagged.

The ground in this area is rocky, but it's covered with a layer of moss. Some moss mounds are soft; others cover boulders. Bark slides off trees, snags catch my kneecaps, branches rip my clothes. I fall several times. I have to stop often for my sanity. As soon as I sit, my head fills with strange images—a Hummer perched on top of the pyramids, a public swim for sharks. I'm awake but my brain is generating the random images found in dreams. My eyes shut and my head drops down for a few seconds before I snap it back up. I walk on.

It takes me eight hours to walk and ribbon less than one-and-a-half kilometres of the fire's edge. I walk out at the end of the day soaking wet from the rainwater hanging on the branches. I'm an overcooked vegetable gone cold. I meet Dan on the line.

"Hey man, is everything all right?" he says.

"Yeah, I'm good."

"You're vibrating."

I look at my arms, which are stretched out holding on to two saplings; they're shaking from my shoulders down to my hands. I hadn't noticed.

DAN HAS SPENT most of this tour away from the crew, working on bigger operational issues, and the crew has mainly been run by the squad bosses. Often Dan has kept Brad at his side to help him. Today I join them to get a look at what's going on elsewhere on the fire. The three of us start by driving an equipment operator named Terry down to his machine. Terry is more talkative and fitter than most equipment operators. A few years ago he bought a five-acre piece of land outside Prince George. The property was surrounded by pines. All of them were killed by the mountain pine beetle epidemic. For a living Terry sits in a machine cutting down hundreds of dead pines a day. Every few weeks he makes it home and looks out at dead pines in his backyard.

The conversation turns to women. One of the other guys driving a Cat on the Superguard has just hooked up with a lady. Terry is happy for him.

"Usually he gets the older ones," he says. "I see there's some nice-looking girls on your crew," he adds. "That Ingrid sure is smiley."

"Yeah, Brad, what do you think, would you say Ingrid is nice?" Dan asks, his voice going high at the end.

It's an easy opportunity for Dan to get a small rise out of Brad, who hates this kind of attention.

"Yeah, she's all right," he says.

End of discussion.

There are lots of guys like Terry in the world, men working in the bush and living in tiny northern towns. They spend decades alone in the cab of a machine, doing mostly the same thing every day. It must be strange, then, to briefly be put to work fighting a fire, and stranger still to have a bunch of beautiful women show up as part of this firefighting effort.

I take a flight in the afternoon. Dan sits in the front of the helicopter, strategizing with management. I sit in the back and stare out the window. The fire stretches as far as I can see. It's so big that different parts of it are receiving different weather. The blacker columns of smoke to the south indicate there was less rain down there. Looking to the north the smoke is blue and thin. The forest below is dabbed with pockets of dead, grey pines among the dark-green spruce and the light-green blooms of deciduous trees.

We fly over the Superguard and see newly exposed tan-coloured dirt and the bright orange and yellow of the equipment dwarfed by its width. Even from the air it looks big. The scale of this fire, incomprehensible from the ground, remains so in flight. Beyond the fire are the glacier-capped mountains of coastal British Columbia. The distant columns of smoke are bent toward the mountains as if heading into battle. The mountains look to lose.

As the spritz of rain recedes into the past, the fire continues to burn aggressively. It's nearly doubled in size since we arrived and is now somewhere around 100,000 hectares, well over two hundred times bigger than Vancouver's Stanley Park. Nobody can be sure, though; the smoke hasn't allowed for an accurate measurement. One certainty is that sometime in the last few days the Chelaslie River fire has become the biggest fire in BC since 1982 and the third largest in the province's recorded history. Fires bigger than the Chelaslie occur in the far north of the country, where the boreal forest grows free of barriers like mountains or roads. But 100,000 hectares is something to be seen at a latitude this far south.

One afternoon management decides to burn off from our Superguard using a helicopter. While this plan was being formulated, we'd been constructing hose trail, two kilometres of it. We'd punched into the woods off the Superguard and were following the ribbon I'd laid out along the fire's edge a few days earlier. We were going to hose it all out, lock it down. Now it will be burned up. Three days of work for nothing. Again.

We were close too, within five hose lengths of completing the trail and getting water on the whole line. It would have taken a few hours at most. There's grumbling when they tell us to leave so the choppers can move in and start burning. Not even the prospect of courtside seats for the coming inferno can please us now. We talk in low voices as we walk the trail one last time.

When we come out of the bush Dan is at the trucks.

"We had that," he says to me as we gather at the edge of the Superguard.

"I know, man," I say and shake my head.

When we're denied an opportunity like this, we sometimes grumble about how much the air side of firefighting operations costs. Almost half the money spent on firefighting in BC—which can be anywhere from $100 million to $400 million depending on the year—goes to air operations. A single helicopter costs much more per day than an entire unit crew.

Before we leave, Tabes and I are asked to haul as much hose and gear off our trail as possible before the burn-off begins. We take two quads, driving through a cutblock to get to our gear, not following any trail, smashing down saplings and tearing over stumps.

I didn't grow up with quads and snowmobiles like a lot of my peers. I'm not too comfortable with them. From what I gather the whole point of owning a quad is to beat the crap out of it, an odd thing to do with a piece of equipment that retails for somewhere around $10,000. Still, there's something to this quaddin'—the few times in my life I've done it, I've felt dangerously happy.

To get the hose out, we tie it to the racks on the quads. We get a couple of loads to the trucks but the fire is hitting another gear. We're running out of time. When we get to the hose trail for one last load we find the fire has leaped across the trail. I grab handfuls of half-burnt hose and throw it into the blaze. I don't know why; perhaps to avoid confusion as to what hose is salvageable and what isn't. Perhaps also because there's something cathartic about putting it out of its misery.

The fire is alive all around me. It's sucking in air and the wind it creates tugs me toward the flames. I try to drag a few lengths out but I don't make it far before stopping. It's too hot. The fire is moving too fast. I drop the hose where the fire is hottest and run through the intense heat back to the quad. I meet Tabes at the Superguard. He appears to be enjoying the moment more than I am; after loading his final length he says, "See ya, Willy, it's been fun."

It's late by the time we get back to the barge. In the distance we can see helicopters circling the fire with torches for burning off hanging underneath the skids. The smoke bomb, now part of the landscape, stretches out for thousands of hectares behind them.

THE FIRE CONTINUES to rage overnight. As it grows the Ministry gets looser with permitting burn-offs, which is what we'll do on our final day.

Tabes and I are support on quads. The burn-off starts slowly and we drive down to the pump site on the lake with the meadow. The ground underneath this meadow would be mucky most of the year, but drought has left it mostly dry.

I reach the pump site but before starting the pumps I take a sharp turn into the untouched field, pressing my thumb into the accelerator. The wind is warm on my face. Tabes and I make a few huge, sweeping turns, hollering nonsense as we steer the quads as close to a collision as we dare.

Manic cackling at high speeds defines the rest of the day. Between delivering gas and managing the pumps, we rally down miles of guard twisting out in every direction. Standing up off the seat, peeling around corners, pelting down straight stretches. By this time tomorrow, I'll likely be at the Smithers library doing some reading. What's happening right now is the antithesis of library time and, given current feelings of invincibility, the only thing I'd do at the library right now is a doughnut on my quad on the front lawn. On one wide stretch of guard we ride side by side, twisting and turning, leaving a cloud of fine dust, a forest fire roaring on all sides.

Moments like these happen only once or twice a year and as leaving this job becomes a possibility, they feel all the more precious. Yes, I could buy my own quad when I'm done firefighting. I could even drive it fast, possibly with a friend who also has a quad. But I likely wouldn't get paid for it, and unless the world descends into *Mad Max*–ian dystopia, I for sure won't do it in the middle of a blistering wildfire.

The burn-off is moderately successful but behind us, where we abandoned our hose trail and the helicopters worked last night, the fire has discovered a pocket of unburned fuel. It's advancing on the Superguard. Suddenly this impenetrable creation looks under-matched. Orange flames erupt through the forest canopy like lava from a volcano. The wind is pushing it all toward the green side of the guard.

Meanwhile, the radio traffic is gathering into a continuous broadcast of panic. Management is trying to account for all crews, watching from helicopters as fire rips through the forest. The wind gets stronger and the sun is out. We'll be flattened if we don't abandon our post and hightail it back to the barge soon.

As calls for us to leave become more urgent, a bunch of us spread out on the hose lay along the Superguard. We spray down a stretch of fire we thought we'd put to bed a few days earlier. Others walk the green side, looking for sparks.

Eventually we make the call to pack up and leave. I walk the line one last time with Dan and we open all the water thieves and aim them at the green side. Something is wrong with the pumps, though; there's no water pressure.

We drive down the road. Dan asks how long I think it will take me to run the nine hundred metres from the road to our pumps, fire them up again and run back.

"Fifteen minutes," I say.

Brad laughs in the back seat. "Wil-ly!" he says, incredulous.

As we pull up to where the road meets the rough trail down to the lake, Dan says, "Let's see what those long legs can do."

Before we stop I've opened the door and I use the

door frame to catapult myself out of the truck and down the trail.

I'm running in full fire gear, hard hat swinging, feet jostling around in my loose boots, chest pack bouncing off my sternum and flailing up into my chin.

On the radio I hear management call Dan. "Ranger One, it's West Div."

"Go ahead."

"Hello, Dan. Just checking to see where your crew is."

Dan lets the silence hang for *so long*.

"On the way out," he replies finally.

He's buying time with his unspecific status.

"You confirm you're driving out now?"

"Yes."

I'm still running. I want this thing inside fifteen minutes.

"Ranger Charlie, it's Ranger One."

"Go ahead."

"Come back to the truck."

Click, click. I press the mic on my radio twice to signal understanding.

Cut short. I run back to the trucks wheezing, the sweat pushing through every dirty pore. My lungs feel shallower than a month ago.

The only sound in the truck is my gasping for air. Dan isn't happy. We had a chance to hold the piece of ground we were on, but we'd been sent off by people who'd only seen it from the air. Who knows, though, maybe it's bad.

We're the last truck to leave the fire. The rest of the crew is waiting for us at a pullout partway back to the barge. I jump in Charlie truck with Tabes, Kara and Brian.

We drive to the barge and I shake off my exaggerated seriousness now that I'm away from Dan.

I don't really care if we lose that bit of work. Anyone who thinks we have a chance of winning this fight is delusional.

When we get to the barge, I put in a huge dip of chewing tobacco. Brian wanders off. I ask the other two what their favourite day of the deployment was. The nicotine buzz pushes my defences down, and our conversation is free of the little formalities I cling to as a squad boss. We all agree the best day was the last day the three of us were burning off. The day we threw the garbage into the fire.

The nicotine is coursing through my system now, putting me at ease. I lean out the truck door and spit onto the road.

The barge ramp scrapes against the south side of Ootsa Lake and we drive onto its ravaged wood deck one last time. The nicotine buzz starts to wear off. I want another dip but nothing will be as good as that first one. I jump out of the truck and lean over the metal railing of the barge.

We've been riding the barge in the pre-dawn light or the afternoon smog but today the sun is out. The rusted cables guiding the barge are dredged from the bottom of the lake, momentarily exposed to the sun. I take out my chew and flick it into the dark waters of the man-made lake. It spreads out into little flakes and recedes into its depths.

WE RETURN TO the base from the Chelaslie River fire on a Tuesday afternoon. Brendan, who did my saw exam, has recently returned from the east side of the fire, and he's the only one around. I duck out of our subdued preparation for the next tour and wander into his office.

"How was it?" he asks.

"Good. Tough but good."

"Yeah, you guys were really getting pushed around on that side, hey?"

"Yeah, I don't know if anything we did had any consequence."

"I don't know if anything anyone is doing is helping right now. These fires are crazy. It's too dry."

I go back to organizing my gear in the locker room. Meanwhile a call comes in for Brendan to get to Houston fast. A strong afternoon wind has just chased two IA crews off a small fire high on a ridge ten kilometres from town.

As we finish paperwork in the meeting room, Brendan gets in his truck and starts driving to Houston.

Leaving the base, I look to the southeast, where Houston sits behind rolling hills and farmland. I see a mushroom cloud of smoke surging into the light-blue sky.

5
China Nose Fire Tour #1

August 17–30, 2014

There are now more than three thousand people fighting fire in the forests of BC, only sixteen hundred or so of them employed by the BC government. We've brought in help from across the country and are now reaching farther: Australian workers will be here in the coming days. Most of the fire activity is in the northern half of the province and the number of hectares burned so far this year is around 300,000, an area bigger than all of Metro Vancouver.

We've parked the trucks at the top of a hill on a recently logged plateau. Our immediate surroundings are scabby and charmless. This is the far eastern end of the China Nose fire, a thirty-five-hundred-hectare burn sitting perfectly halfway between the towns of Houston and Burns Lake.

Yesterday this fire sent the region into temporary hysteria. A warning on the Village of Telkwa Facebook page suggested the fire was going to cross the highway and cut power to the region. Locals rushed to the gas station and grocery store for supplies. This reaction was probably more a result of seeing too many zombie shows than any real threat. From the plateau we get a sense of that "threat." It's even more implausible than we expected. The fire is a long way from the highway, and the land in between—marked by farms and deciduous trees—would make for poor burning.

The fatigue hits right away, like we never had a break. The difference between the last deployment and this one, our third two-week stint in a row, is profound. I felt fresh for about five days on the last tour. This time I'm zapped just from hearing Dan's initial instructions. I have a hard time getting back into it with Dan. After being frustrated with him last tour, I'm slow to meet his eye.

The day is made tolerable by a series of quick uppers—chips, caffeine and candy. All leave the guts feeling rancid but keep your corpse animated.

Our first job is to put out a spot fire that ran away from the main burn the day before. I cut most of the day, falling burning trees. My feet sweat from the heat of the smouldering ground fire. The exhaust from the saw fans the hot coals embedded in the tree trunks. Fire springs from the coals, and the smoke and heat prevent me from making accurate cuts.

THOUGH THE FIRE is four days old, there's no fire camp set up yet. Instead of staying at a local hotel or driving back to the base, Dan sets us up at Jon's place.

Jon is an ex-Ranger and devout Dutch Christian. There's a large Dutch community in this area and although they've been here for generations, they've done a lot of their regenerating within their community. This has given the northwest a steady supply of tall, blond, beautiful men and women. We'll catch a glimpse of them once or twice as teenagers while playing against their Christian school sports teams; after that they disappear for a half-decade or so and re-emerge with four children and a large pickup truck.

Jon is one such Dutchman—tall, blondish, skittish and hard-working to infinity. He's homesteading just south of Houston. His driveway connects with the main logging road coming into the Houston sawmill. We turn our trucks off the muck of the logging road, watered to keep the dust down, and up his driveway, where grass grows between two tire tracks. The flakes of mud fade out and we enter pastoral farmland. His house is earth-coloured. It sits in the lee of a small mountain with high meadows just craggy and secluded enough to keep mountain goats. We're camping in the green field adjacent to Jon's house and as we drive by his house, he bounds off the front deck carrying a cooler.

"I got you guys some corn!" Jon says. His exuberance is his trademark.

"Awesome, man," I say, and I mean it.

We talk briefly about work before Jon shoos us out to our campsite. "You guys must be tired. Get yourselves set up out there and get some sleep."

The caravan of trucks moves into the field at a walking pace. We're heading to a pile of gear dropped off earlier in the day. We're camping on the flattest part of the field, a low corner next to green timber. The twilit mountain stands above us. Across the field an ember of light glows in Jon's living room window.

We park at our pioneer camp and slump out of the trucks, descending on a metre-high stack of pizzas we bought in town. I open Jon's cooler and find a stack of yellow corn. We lounge in the pasture among the anthills and cow patties, eating our corn and trading slices of pizza, chewing with the zeal of calves treated to fresh silage. The stars are out before we're done and we set up our tents in the dark.

MY SAW PANTS are damp with yesterday's sweat when I put them on in the morning. A white film covers part of them. It could be salt—or mould.

Cutting continues today. We're working in a stand of spruce trees as big and healthy as you'll find anywhere in the Interior. Fire burns their roots but rarely climbs the tree; it's deceptive. Spruce trees can look fine even when their root systems are completely burnt out. When this happens the wind can blow them over and they won't make a sound as they fall—there's nothing left to snap.

I put cuts in a tree until the exhaust from my saw fans the flames. When this happens I step back into a patch of green moss, take a breath, check my surroundings and let my feet cool.

In the afternoon Tabes and I are sent to Houston to buy groceries. We're going to camp at Jon's again tonight. We head to the Houston Super Valu; it's located in the Houston Mall, whose sole tenant appears to be the Houston Super Valu.

Everyone in Super Valu wants to talk to us. Locals thank us and ask how it is out there. We're diplomatic with our answers and Tabes, unexpectedly, is a natural at this. We're not too positive or negative. Talk lots, say nothing. People seem satisfied with our answers.

For years we've swept into grocery stores and never had to answer a question. The fires we fought were miles beyond any living soul's field of vision. This is a nice change; we do so much meaningless stuff that the attention feels good.

We get back to Jon's at dusk with $445 worth of pitas, chicken wings and pop.

Dan rolls in at 10:30 after attending a day-end meeting. I walk up to the door of his truck.

"My wife's here," he says.

"Oh man, that's great," I say.

I think about this for a minute and decide I would pay $1,000 for that option right now.

We finish our pitas and chicken wings, our mouths coated with the savoury sauces hiding shoddy ingredients. All of us retire silently to our tents. All except Addison, who elects to fall asleep on the ground.

From the tent I hear the wind in the aspens. In the distance there's howling, wolves or coyotes. A timeless feeling permeates this chunk of property with its dark soils and loud leaves. Timeless and empty. We live in a world that champions the majesty of nature. We're told time in nature is restorative; it helps us recharge and discover meaning. Sometimes, though, I've found it has the opposite effect. Some nights it feels as though I could unzip the tent, look out and find I'd spent my life dreaming of civilization. Some nights all the sounds, or all the silence, tell an unsettling truth—we're alone and insignificant.

IN THE MORNING we stand around the trucks eating Froot Loops and granola and handfuls of blueberries. It's spitting rain as we leave for town. We stop for coffee at the Houston A&W. Inside the restaurant, a horde of usuals sits around drinking coffee. They ask us the occasional question. The people who ask, mostly denim-clad men in their sixties, do so in a performative way, showing everyone their knowledge of both the local terrain and forest fires. The scene is too bubbly for six in the morning, but I shouldn't be surprised; the entire town of Houston goes to bed at eight p.m. and gets up at four a.m.

Partway through the day, the four of us in Charlie truck are called to another part of the fire to do some danger-tree falling. On the way, we see China Nose Mountain, for which the fire is named. There's some disagreement over the correct spelling of the name. During the building of the railroad, Chinese workers were said to have found gold in the creeks below the mountain. White settlers would use the phrase "The Chinaman knows" when discussing this. The provincial park surrounding the mountain is called China Nose, but a nearby ranch is called China Knows. Either way, it's an impressive slab of rock. Formed by volcanic activity fifty million years ago, it's the same vintage as the more famous features of Yellowstone National Park. A four-hundred-metre sheer cliff dominates one side of the mountain. Balsam and spruce trees butt up against the bottom of the rock face, at risk of being destroyed by the boulders that crack loose during the spring thaw every year.

Charlie truck rattles up the rough road toward China Nose, where we'll be working this afternoon. Waiting for us at a junction under the cliffs are Brendan and two Australian workers.

We get out of the trucks and shake hands. The Aussies are dressed in electric green jumpsuits that appear to be made out of Kevlar. I shake hands with a man named Ian, who's of medium build and middle-aged, but telling them apart is difficult due to the shocking green onesies they're all wearing.

"How ya doin' there," he says in an accent strong enough to put more fissures in the cliffs of China Nose.

The first item on the agenda is making fun of the phallic nature of the zoom lens on my camera, which I have close by to take some pictures of China Nose. I'm excited to meet these guys and pleased with how well I engage with the coarse banter that is their national dish. We end the meeting with jokes about getting down to business and not slacking off.

Brendan, who's been showing the Aussies around, wants Tabes and me to fall burnt trees near the road, making sure it remains clear for crews to access the fire. Kara and Brian control traffic while Tabes and I work our way along the road. The undercuts we put in the trees look like toothless smiles, ill-fitting for such proud and imposing plants. There's something noble about how these trees fall, at least when cut with a chainsaw. The way they first shiver, then topple, slow and elegant.

I want to be efficient like an industry faller, to move through the woods with my saw like an assassin. I'll leave no sign of struggle, no broken limbs at the tops of other trees, no extra cuts to get the tree to lie on the ground. Only a clean mass murder of trees.

It doesn't go anything like that, though. It's sweaty

work and trees hang up in other trees and I get scared. As a kid I was the first to run away when there was a hint of trouble. I'm the same way now. My flighty nature is at its height on these falling days.

A FIRE CAMP is set up in a field on the side of the highway near Houston. It's next to the train tracks and, unlike our previous camps, it has the air of a festival about to begin. This probably has something to do with its proximity to town, cell coverage and the freeing sensation of being in an open field rather than a cramped gravel pit. New crews arrive clean and smiling. They play hacky sack or football, or catch up with old friends.

With all the new crews arriving there's a change-up in tasks. We're moved to a part of the fire burning across a gorge. We've had machines build guard for us all summer; today we'll have to cut a fuel free down the steep pitch to connect two guards.

While a big group gets to work falling, Tabes and I are told to find a water source above the fuel free. Dan tells us to "be tenacious" and we are. But we don't find anything. We're searching on the dried-out top of a wide, barren plateau. It's a recent logging block and all that's left are a few of the brittle bones of an unremarkable stand of trees. It's obvious the only water is at the bottom of the canyon.

In the late afternoon my squad helps Rob's squad, which is hosing a small escape a short drive from the steep canyon. On arrival, everyone piles out of the truck, eager to help. One of Rob's squad members is hosing an old slash pile just off the road. Apparently she's been stewing on the hose all day, tired and bored. She spits poison at people the first chance she gets, saying there's no room for us and the water pressure is no good anyway. She's just one person, but when I come upon the group they're cowering back on the road, giving her a wide berth as if there's a bear in that part of the bush.

The testiness comes out later as well. During the day we messed with Warren's truck radio, switching all his presets to Christian rock stations. Then, when we were driving down to meet Dan for a meeting, Tabes cut Warren off in order to be the lead in the convoy.

We gather in a circle for the meeting and go through what's been done today and what we'll do tomorrow. Dan asks Warren about his line-locating progress, which has been slow.

"You're still not done?" Dan says.

Warren's eyes narrow.

"It's a long ways to go," he says.

The meeting ends and the circle starts to collapse when Warren speaks up again. "One more thing—quit being assholes in the trucks."

Everyone freezes. Warren hardly talks in groups, never mind loudly and with light cussing. After a long silence Dan tries to defuse the situation, asking what he means. Warren's voice shakes as he explains being cut off by Tabes. Tabes looks surprised, not realizing he'd bristled the ten-year veteran of the Rangers, a veteran whose climb through the ranks to a respected position was long.

By the time he became squad boss the previous summer, the only person on the Rangers who remembered Warren at age eighteen, with a full head of hair wrapped in a white bandana, was me. There's usually a special affinity between people who start this job at or around the same time. But with Warren and me, it felt like a rivalry, at least on my end. This feeling didn't go away until the summer of 2010, our fifth on the crew together. That year was similar to this one; we fought a lot of fire close to the base in Telkwa. A few of us set up tents at the base and slept there after work. Warren and I were the only regulars. We stayed at the base for about a month. The two of us would eat canned soup for dinner and cold cereal for breakfast. We'd go to bed in the dark and get up in the dark. We had a boom box and the *Top Gun* soundtrack on CD. The days were hectic, but the evenings at the base were idyllic, every night a bit of listless relaxation before going to sleep.

Five summers earlier Warren and I had decided we didn't like each other and hadn't bothered to check if that had changed until 2010. Why would we? The check was forced upon us by circumstance, and it turned out that we'd changed a lot, as people tend to do between their teens and mid-twenties.

Maybe by cutting him off in the truck Tabes had triggered some sort of flashback for Warren, something that reminded him of the days when he was at the bottom of the unit crew pyramid.

THE FUEL FREE curves as it heads downhill, giving the pleasing impression of a fresh-cut ski run without the snow. The entire crew is here to finish off this project. There are a dozen of us ready to swamp. The rest are running saw, bucking wood into liftable chunks.

In one big group, we get to work stacking the wood up high on the green side of the fuel free. It becomes a collage of rounds and branches and decaying chunks of wood. A reserve of goodwill is tapped here; the suffering through workouts and tedium has helped us arrive at this point.

When we're finished swamping, we dig guard. The aches arrive soon after we start. My hamstrings tighten, my neck gets sore, my wrists feel like rusty hinges. We build the guard, all edges trimmed, uniform depth, perfectly centred on the fuel free. It looks like the answer to a math problem.

The day is getting boring. But then a mystery voice on the radio asks if crews on the ground can try out their drip torches to see if anything will light.

We take "Try out your drip torches" to mean "Burn this mother down," and that's what we do.

The moss is wet from last night's sub-zero temperatures, but despite the damp ground, all other fuel is the driest it's been in years. We fight through the thick forest as trees candle around us, some going up in a flash so quick that it feels more like minesweeping than burning off.

I walk deeper into the bush and empty the dregs of my torch at the bottom of the biggest spruce I can find. I stand right underneath it and watch. The smoke off the top smears the sky with wet brush strokes. Little branches and bits of old man's beard float down glowing orange, hitting the broad-leafed understorey, searing the leaves with little black spots.

The guard gets steeper and eventually it's too steep to have everyone burning off. Tabes and Brian end up burning all the way to the bottom. Our Nomex clothing will repel fire, but only up to a certain point, and partway down the hill Tabes's pants catch on fire. He drags one leg along the ground while slapping at it with his hands. With the leg fire out he continues with the burn-off. He crosses sections so steep his feet dangle underneath him and he hangs with one hand from brush or branches. At the bottom he stumbles back to the fuel free. His face is flushed, glistening. He collapses onto his knees and puts his head into the cool moss of the creek draw.

After a minute of rest he looks at me. "You got any water?" he asks, like it's his dying wish.

I don't and he has to climb back out of the creek draw to get to his pack. It's hard to watch him struggle up our fuel free, the hill orange and roaring on one side, green and quiet on the other.

THE NEXT DAY, Tabes and I take another crack at burning up a bit of land that didn't take off in yesterday's deluge of drip torching. We fall a few trees to encourage a ground fire and then bring out the torches for another attempt. Rob and Warren are nearby to help.

The little section refuses to succumb. I take the lid off my torch and dump the last few drops of fuel under a thick bed of branches. I light the branches but realize too late that the torch in my opposite hand is hovering directly over the flame. The flames leap into the torch and create a cauldron of fire. Instead of burying it in the dirt I put my face over it and blow. The flames leap out of the can and singe half my facial hair. My beard has been growing for three months and it writhes in the heat, giving off that awful stench.

Shocked, I look up to see if anybody saw my idiot move. Standing nearby is Rob, who is leaning forward, mouth open, eyes wide, ready to roar with laughter but waiting until he knows I'm not hurt. Our eyes connect and I indicate I'm okay, not by smiling but by sending him the same open-mouthed look of shock. A second after that, he collapses forward and howls with laughter.

He keeps repeating, "The look on your face! The look on your face!"

With all the falling trees and burning off I've done lately, I'd been feeling pretty cool. This is a reminder—I'm not cool. And I'm lucky it wasn't worse. In the following days, word of my burnt beard spreads through camp. People on our crew and other crews come up close to my face, see the curdled beard with its consistency of frayed rope, and cackle away.

A THUNDERSTORM APPROACHES. Little pockets of fire, previously resting in the moss, are brought to life by the strong winds preceding the downpour. The flames look tenacious under the dark clouds. I stand and watch a pocket of heat thrash around as it starts spitting rain. The flames, billowing and wide on the forest floor, find a spruce tree. One more kill before they lose the battle to the rain.

The fire grabs on to the spruce and its lower limbs begin to sizzle. Flames climb the tree and when they reach the top they balloon out, feasting on the fat clump of recent growth crowding the tree's peak. The fire burns slow up there and the rest of the tree cools down and fades to a less intense orange. But the top still burns bright, the colour of a street light against the storm.

The rain sedates the fire. To us, it's a tonic. It tells us we can dial back the intensity.

Most of the last three deployments have been spent

chasing fire. Some years we spend only a few days on a fire's tail before we gain the upper hand. In busier years we might spend a few weeks at it. But six weeks is a long time to keep your guard up.

Between the rainfall and the time of year, we intuitively know the pace is about to change, even if briefly. Into the chasm stress has left behind comes a torrent of laughter. On the drive to camp we laugh about the overweight medic assigned to our section of the fire, who looks like he would collapse from overexertion if he were ever called to help us. We laugh when we get a flat tire on the way out. Our tires are brand new, advertised as rugged bush tires. We bought them because the crew had twenty-four flats on our last fire. We laugh hardest when, after changing the tire, we lower the weight of the truck onto our fresh-from-the-shop spare and find that it too is flat.

After dinner, in an open patch of grass near the dinner tents, Brad and I talk with a few ex-Rangers who now work on a unit crew out of Kamloops. All have been friends with Brad since childhood. We're telling stories, bouncing back and forth with half-truths and mudslinging. It's getting dark and cooling off. We lean in and laugh hard. Our laughter transitions into coughing from the smoke in our lungs and the sickness worming its way into our systems with the arrival of colder nights.

One of a trio of friends has moved on from firefighting. He's now an apprentice electrician working on a dam in the far north of BC.

"How's Myles doing?" I ask. "Do you think he misses the job?"

"Oh yeah," someone says. "Everyone misses firefighting."

Later on, I'm at the trucks on my own. Ingrid is in the front seat of the next truck over. By the light of her phone, I see her silhouette, her hair in a tight ponytail. She's talking on the phone and her face is rigid.

When she's done talking she stops at my window. The tautness of her face matches the tone of her voice. "Hi, Willy."

"Hey, what's going on?"

"Oh...I don't know."

"What do you mean?" I say, laughing.

"I never thought I'd take three years off," she says.

She's talking about the three years that have passed since she finished her undergrad degree. Ingrid is the offspring of wealthy Vancouverites. She's part of a different economic class than most of us on the crew. She's living a life miles from what's expected of her right now.

"I feel like I'm going somewhere, but I'm going nowhere at the same time," she says. "I don't want to tell my mom I'm an EI bum."

The wildfire branch of the Forest Service is built on EI bums, but Ingrid isn't part of this foundation. Her life is supposed to be going somewhere. In her world, standards must be met. Quotas of enriching experiences must be filled and verified. This job is one such enriching experience, but it has outlived its use.

I'm sympathetic to her problems. Maybe my meeting with the guys tonight has something to do with it. For those guys tonight, especially her boyfriend, Brad, there are no expectations from outside, only the ones they place on themselves. They've collected EI, no problem. They've worked worse jobs and they may never work a better one than firefighting. In the winter they're roofers and oil patch monkeys and general labourers. There was such freedom in the way they were talking and laughing.

Everyone misses firefighting.

A HELICOPTER LANDS, then sits lifeless on a pullout at the side of the road. Inside are Brendan, an Aussie and the pilot, who is writing in his logbook. A few minutes earlier I was writing as well. A few crew members have to get back to school at the end of this tour. Before they leave I must write evaluations of their performance this season. But more pressing than writing performance reviews is taking a flight over some new ground we'll be working tomorrow.

Dan and I climb in and the silence in this normally deafening machine is odd. The pilot finishes writing and starts flipping switches and pushing buttons, bringing back the noise. Soon we're throwing bits of debris across the landscape as we climb into the air.

In no time I'm sick to my stomach and exhausted from the succession of long days. We're twisting and turning in the air as the sun beats through the plastic windows.

I can't sit back on this ride, though; we're scouting new territory. We're heading to a remote section of the fire with no roads and no water. We fly around looking for potential helipads and possible water sources. Dan and I make a plan and then we're shuttled back to our spot at the top of the fuel free, where the fire is quiet and the rest of the crew is hosing, bored to death at the end of a nozzle.

THE NEXT DAY we pack the back seat of a bigger helicopter. This particular Bell 205, now blue and silver, was once painted camo; it's a veteran of the Vietnam War. Images and movies related to the Vietnam War are pretty common in popular culture. This becomes apparent when you're crouched in a hurricane of rotor wash as the chopper approaches to pick you up and "All Along the Watchtower" starts running through your head.

Our pilot, a smiley young guy from down south, does his part to glorify the job of being a pilot. He swoops through the valleys and over the ridges, nose-diving off a cliff edge and making all of us scream with joy in the back. He doesn't just do it for us; one day I see him do his signature nose-dive off the ridge.

I turn to Warren. "Does he have any passengers?"

"I don't think so," he says.

I move hose with Dan most of the first day at our fly-in spot. For the first time this summer, we're on a mountainside. We've been in steep sections at times this year, but not in the relentless, obstinate clamp of a mountain, where leg muscles or lack thereof take on the starring role in your physiology. Every potential walk down the hose trail is calculated to make sure it's necessary. (It always, always is.)

We visit as we walk. Dan asks me about my plans for the winter. I'm not sure how closely he's listening, but I talk anyway.

"Well, I have a family reunion at my grandma's place in Belize, but that's about all I really have to do."

"What?"

"Yeah man, my grandma lives down there in the winter so we're doing it there."

"I should bring the family down there!"

Dan is radiating genuine excitement. He for sure will not do this. He's just fantasizing about having fun with his family instead of rolling in the unit crew muck.

"Look at this, I'm tearing up," he says.

He's half yelling this and I look at his eyes. They're misty.

At the top of the hill we come out along a ridge used by mountain goats and some little birds whose only evolutionary skill seems to be grabbing updrafts and rocketing straight up.

I've been around Dan a lot the last two days and it's been good. He seems calmer and I can take more of a back seat. This whole tour has been good in that regard—less responsibility, more time to just be a worker.

At the end of the day we meet the rest of the crew at the helipad. I sit down with Pete.

"How you doin', Willy? Excited for some days off?" he says.

He's looking forward to going back to school. The first domino to fall.

After scoffing I say, "Not yet, man."

I feel bad for being dismissive of Pete. But for some of us, there's still lots of season left.

IN THE GAME of "bush" everyone gets in a circle. We each have three small sticks in one hand. Each person decides how many sticks they'll keep in their hand before thrusting a closed fist into the centre of the circle. We then guess how many sticks are in the circle. The right guess wins.

Today is our first full day of patrol and our first game

of bush. The game determines who gets what spot on the line. We patrol in groups of ten, looking for any leftover heat. Certain spots in the line are more desirable than others. Being in the "one spot" is often least desirable. You walk the fire's edge, where burnt meets unburnt, and it can be hard to follow.

Not today, though: the fire's edge also happens to be a cliff edge. It will be easy to follow and will also have a great view. I guess the right number early and take the one spot.

Patrol starts. We're on steep ground. Our boots do all they can to grip the hills that slough away with ease after their recent trauma. We cling to branches, burnt or rotten, as we make our way down a ravine. We crawl up the other side on all fours. Gas cans and small pumps are heaved in front of us. We nestle them in any little relief in the steepness and scramble up after them, arriving at the top of each draw breathing hard. There are a few hot spots in our cliffside travels. We grub away and the ash rises thick from the ground.

IT'S GREY, COLD and windy as patrol continues the next morning.

But by the end of the day, when we get to the helipad, the sun is shining. Our pad is at the top of a steep hill; it's a jagged piece of rock with sharp bits of shale and hardy alpine brush so gnarled you can't even sink an axe into it. While waiting for a ride out, some crew members chuck two rookies into the water bladder for the hose lay we'd set up. The rookies go in, laughing and shouting, a whirling, fighting mess of legs and arms.

After one rookie, Nate, gets tossed in, he continues

to lip off Brad. Nate is known for accidentally hitting his truck's power window button and asking the rest of his truck who was doing it as his head became jammed between the window and the frame.

Both Brad and Nate played junior hockey. During the season they've spoken bits of their native tongue to each other. They say things like, "He's filthy" (good) or "He's dusty" (bad) to describe different people they've played with.

I can't hear what Nate is saying to Brad, but eventually Brad has had enough. He knocks off his hard hat and walks toward Nate. He throws down his work gloves as he's walking; there's still a bit of a smile on his face. They start wrestling on the helipad.

I know Brad will win before it starts. He has too much to lose. Plus he's heavier and has an older brother who used to do things like tie him to his bunk bed ladder, tickle him and then leave him there. Brad scores high on all criteria for unit crew success—toughness, humour, loyalty. In a torrent of grunts and ripping fabric, Brad takes Nate down and reasserts himself as the alpha male on the crew.

On the sidelines, the crew jokes and eats snacks and comments on the match like they're half drunk at some stage play. I look at Ingrid. She's standing near Brad, who's half soaked with water, his pants ripped to shreds and his shirt untucked. He's smiling as the excitement of the match dies away. Ingrid's not talking to any one person, but she looks vaguely pleased. She's half turned toward Brad and is absentmindedly fiddling with something in her breast pocket.

Warren and I sit down away from the group. He's

stressed after a day of writing evaluations. As we sit there, Warren takes out his pipe and we pass it back and forth looking at the distant rain squalls and farms below the mountains. In the background, the banter of the group is like a triumphant chord—the sun is shining, the hierarchy has been reinforced, some are heading back to school soon.

Warren is leaving for a funeral tomorrow. Despite the sombre occasion, we're all happy he's getting a break. By contrast, a month ago Chris missed a few days of work to attend a wedding and we harped on him. Early in the season every fire day counts. We thought Chris was out of his mind to turn down the money.

Somewhere along the way, taking time off has become desirable instead of idiotic. I love the idea of Warren getting a few days off.

THE NIGHT BEFORE we leave fire camp I fall into a light sleep and dream of firefighting tasks. In the thick of work, dreams turn into grainy reruns of my day. I sleep like this for half an hour before being woken up by Pete and Chris, who come through the door of the tent and the navy light of the outside. People around me snore; Pete and Chris finish their conversation in a whisper. I can't get back to sleep so I go over to the food tent and get a bowl of Frosted Flakes and eat it while watching Jimmy Fallon's monologue on a TV that's been set up in one of the tents. I go back to bed and a few freight trains later I'm up again.

We get back to the base. Some people clean out their lockers, others prep for the next deployment.

As Dan and I are working in the office Brian comes in to say goodbye, a little misty-eyed. The YEP comes in as well. He's a different kid than he was two months ago. He shakes my hand and though he still carries a sneer, he sounds genuine when he wishes me farewell.

Again I'm the last one at the base. The wind moves through the buildings carrying little flecks of dead grass from our lawn. Dusk arrives earlier now and it's getting dark by the time I leave.

From the gates of the base I phone Sue and we talk about everything—music and books and people on the crew and her friends and my friends. My voice bounces off the buildings and carries into the pine flats. At last I'm free to talk loudly with no time limit. As we're talking I walk around the edges of the base, where bits of an old running track still exist and cribbed fences made of pine are in the process of disappearing into mulch. By the time I say goodbye only the faintest trace of dusk sits above the mountains.

6

China Nose Fire Tour #2, Chelaslie River Fire Tour #2

September 3–16, 2014

The tension doesn't rise in my chest as we load the gear into the trucks. I don't go over any mental checklists of what to bring and I don't worry about having spare protein bars or extra socks. It's our fourth deployment in a row. As long as I'm where I'm supposed to be at the right time, it doesn't matter if my mouth is agape and my eyes are half shut. I only need to be semi-conscious to do the work.

We're ready to go ten minutes early so we stand waiting near the trucks. The guys and the girls are separated—three girls on one side and nine guys on the other. It's like an elementary school dance. The conversation is hesitant, almost awkward. I stand with the guys and we're all waiting for somebody to break the silence, to say something we can agree with or ridicule.

We're waiting to get back to the fire line mostly, where

we can talk about work and make dick jokes and laugh at missteps. Our ability to follow the social norms of town life has deteriorated. Our entire relationship is now built on the crude honesty of the bush.

Somehow, they've sent us back to the China Nose fire. We get to the fire camp and it's cold, the sun still feeling its way through the autumn fog. We wait half an hour for instructions; during the wait we walk over to the kitchen and grab bad coffee and treats. Frost clings to the roads with new-found defiance as we drive high into the woods to a section of the fire that needs patrolling.

We slip into patrol at eleven a.m. After walking a ways I get curious about the time, estimating two hours have passed and maybe I can justify a snack. I look at my phone and it's 11:24.

We're walking through cold cutblocks in clean clothes with a real chance we won't see any smoke all day.

We do find heat, though. When we stop to work, the sound of hand tools clinking against rock travels fast across the land, uninhibited by stifling temperatures. Distant conversations sound close as well. I feel I could pick up a stick, throw it at the blue sky and shatter the words as they fly through the air.

Rob comes by with a plan. "Willy, I think it might be a good idea to move on down the line and see if we can't get some water on some more spots."

I'm slow to engage with him. "I don't know, man. I think there's enough to do here."

I do this partly out of fatigue, but also because of our new sleeping arrangements on this deployment. As long as we're working on China Nose, we're going to commute back to the base each night and go home to our own lives.

Your mindset changes as soon as you know you can get away from the group for the night. If I were staying in the same wall tent as Rob, maybe I would have gone along with his suggestion.

We gossip in the truck on the way home—about who annoys us, who flirts too much, authority figures we've grown to dislike. It's just Kara, Tabes and me, no more Brian. Maybe Brian, in all his rookie innocence, made us less cynical. Now the talk has an ugly fervour, but gossip makes the days bearable.

The patrol continues on our second day. As a training exercise, Warren is acting as crew supervisor. In the afternoon I ask him what's going on in the rest of the province, how much fire activity is left and what our potential is for getting deployed again.

"Not good," he says. "Things are really quieting down—we'll be lucky to get the weekend in."

My face is flash-frozen in an expression of horror. We were supposed to get another full tour. Firefighting is gambling; I'm in way too deep and about to lose big.

At the end of our first tour I worried that we wouldn't get out the door again but we did, and then we did again. By the third tour, management was talking about fighting fire "till the snow flies."

This information sent the fire/money anxiety quadrant of my brain into a rare dormant state.

Now Warren has woken me out of a dead sleep with an air horn. One more day? I slink back into the bush to help with the hosing and brood.

I worry about the possibility of this being my last year. What will I do without firefighting? How will I make money elsewhere? At the same time, the above-mentioned gambling aspect of the job is one of its difficulties—the money is inconsistent. I worry about my lack of hard skills, the ones that give you high starting wages and job security. This in turn causes anguish over not being chosen as a crew supervisor this year, something I haven't thought of in months.

I put in a dip of tobacco and feel worse—weak in the knees. I escape deeper into the forest for a few minutes but then collect myself, because I am supposed to find out if we can get any more hose.

On my mission I walk up the road to where Dan and the other managers are meeting. The guy who was picked over me for the crew supervisor job is there. Silently I insert myself into their group. Usually I'm an insider in these scenarios, but right now I'm a dark smudge at the edge of their managerial glow. I ask a couple of operational questions and go back to work.

I find Brad and a few others tidying up some gear before moving to a new spot. Brad has taken a vow of silence this afternoon in mock protest of the teasing he's getting from the crew. People laugh as they hit him with volleys of insults, to which he says nothing.

After bundling up some gear, he and I go on a mission to find more hose. As we leave the group, the crew gets in a few parting shots, saying they don't believe he'll stay silent once he's out of taunting range.

We get a ways up the road and, staring straight ahead as if staying true to his plan, Brad starts talking through pursed lips.

"Don't look at me," he mutters.

"Yeah man, no problem, that's awesome," I say.

Still without making eye contact, we start talking. I

ask him who has the most unfulfilled hockey potential on the crew. He ponders this for some time. I jokingly tell him I want it to be me. I'm not joking, though—I do want it to be me.

"Uhhh, that's a tough one. Probably Blaine," he says.

It's always the people with strong legs.

I tell Brad I've been having anxiety about leaving firefighting and he responds by saying that his vow of silence has had the same effect on him. Not talking to the others while doing the same menial task for hours and listening to their punch-drunk late-season banter has given him space to consider his own state of affairs. He aspires to work as a lineman for BC Hydro. A good job, well paid. But not as fun as firefighting.

At the end-of-day meeting before we drive home, Dan reaffirms Warren's words: fire season may be over soon. He adds that he doesn't want to "suckle the teat of Romulus." According to Warren, our resident expert on Greek history, Dan's reference is out of whack.

Whatever. We know what he means—we don't want to keep coming out and pretending to work when there's actually nothing to do. Dan talks about it having been a good season and when he catches my eye, as he often does in these meetings, especially the ones carrying hard truths, it takes some effort to nod in agreement.

We get in the truck and I turn to the back seat. "I need some depression chips."

I eat until I'm sick of them. "I need some depression Twizzlers."

Twizzlers also fail to comfort. Tabes and Kara say they're okay with us being done for the year; their maturity further exposes me for the whiner I am.

As we pull into the base, Dan comes on the radio. "Before we take off, gang, I'm going to hook the trailer up and we'll throw the camping gear on. It looks like we're going back to Chelaslie River, either tomorrow afternoon or on Saturday."

He says a bunch of other stuff too but I've stopped paying attention. There are all sorts of alarms going off in my head. Where did the information come from? What if this is rescinded? Still, I high-five Kara and Tabes, my emotions uncontainable.

We park and start cleaning out the trucks. While we clean, a little plastic bit pops off one of the million compartments that decorate our truck. It's small, black and teardrop-shaped. We try to figure out what it is but can't find a home for it.

Eventually we chuck it in one of the cupholders and decide to call it "the Teat of Romulus."

TO ADD A touch of civility to our lives, we sometimes listen to CBC radio in Charlie truck. The program we get in the morning is called *Daybreak North*. It's broadcast to two-thirds of British Columbia's land mass and less than one-tenth of its population.

As Tabes drives us back to Chelaslie River, Kara phones in to see if she can play their morning trivia game. She is selected to go on the air. Kara correctly answers the question: True or false, China is the country most visited by Canadian citizens? (Answer: False, it's the United States.) Then she talks to the host on-air for a bit. Kara is nervous but on the other end the host sounds thrilled to be speaking to a firefighter on her way to the very fires the station has been reporting on for months.

Throughout the conversation, Tabes and I have the radio on, quietly so Kara won't hear herself. We've also alerted the other trucks to our situation.

After she hangs up there's celebratory hollering. I unbuckle my seatbelt and turn around in my seat to give her multiple double high-fives. A few minutes later the station announces the week's winner of the coveted *Daybreak North* mug. It's Kara. We triple up on the shouting to the point where there are a few seconds when we're all going "Ahhhhh" at max volume in perfect harmony.

WHILE WE WERE at China Nose, the Chelaslie River fire continued to grow. When rain soaked China Nose, it only brushed against the mammoth inferno burning in the heart of BC for fifty-nine days and counting.

Fire camp is on the east side of the Chelaslie fire, down the same road we drove to get to the Euchiniko Lake fire months ago. It's set up in green grazing land for a nearby ranch. We drop off our gear and head farther into the bush to hit the fire line for a few hours. On our way we drive by a group of Atco trailers in a clearing on the side of the road. I recognize the spot right away—it's where we stayed on my first-ever deployment.

In my first year on the Rangers, rookies had to peel logs when we were on the base. It was for a fencing project. Warren and I, along with another rookie who's no longer on the crew, spent hot days behind the crew's bathrooms at the Telkwa base, taking the bark off logs and listening to the local radio station. Veterans would come by, use the bathroom, then come out and lean against the railing. They'd stand there and chastise us, thoroughly and in great detail.

They were joking, of course, and Graeme and Warren, because they'd been YEPs on the crew the year before, could take it in stride. I followed their lead and tried not to let it bother me.

One afternoon I was working alone when a veteran came back to the log peeling station.

She stood watching me for a while before I looked up and noticed her standing there. I don't remember her exact words, something like, "Hey, new guy. Fire call, get your stuff."

Because no practical joke was off limits, I was skeptical. "Don't bullshit me. Is this some joke you guys are playing?"

"No, c'mon, grab your gear."

I remember hitting the line that first morning excited to see a fire, but we were working in a cold cutblock. There was the odd patch of blue smoke but mostly the land was just steamed and blackened, like a barbecue after a grease fire.

Following my first fire day, we were sitting around the dinner table at camp. I made a comment about how impressed I was with the well-worn game trails we had seen in the bush that day. Coming from the coast I'd never seen anything like it. This was met with silence. After an excruciating amount of time one of the vets looked up from his meal.

"Where are you going with this?" he asked.

There were a few snickers from some of the other older guys, and I looked down at the floor, humiliated beyond belief.

I share this story with Tabes and Kara and its coarseness shocks them. There's no way that sort of hazing would happen now; the crew is far more inclusive. Of the people on the Telkwa Rangers in 2014, Warren and I are the only ones who worked that fire in late May of 2006.

Our drive to the fire line takes longer than we expect and we only have time to survey the land before turning around and heading back to camp.

In the evening after dinner I'm sitting in our truck, listening to music and taking notes on the day. Dan opens the door and climbs in beside me. I'm happy to have a moment with him. I ramble on about what I'm going to do this fall and next year; my thoughts come out broken and unfiltered. I say I'm afraid of coming changes, worried about the idea of leaving this job.

Dan cuts in. "You have to figure out what you're doing, Willy."

It's harsh, but his words come from a place of sympathy. It's being in flux that's killing me. I suspect there is indecision churning in him too. Dan applied for a new job within the Ministry before this season started. He was also narrowly defeated by a seemingly less qualified candidate. In this matter, we stand side by side. But he's less sentimental than me, and it appears he's further along in sorting out his problems.

He leaves the truck and I go back to writing.

WE'VE INHERITED A mess of a piece of land. The danger-tree falling was sloppy and there's a horrendous four-kilometre-long failed burn-off. There's also a spot where the fire spilled over the guard. It's the firefighting equivalent of a big-city garbage strike. This kind of thing will happen when you've had four cycles of crews from around the world working the same fire.

We're not bothered, though. There's ground to be worked.

Tabes and I go to work falling. We're working in a forest made up mostly of pine. Fire tends to burn a ring around the bottom of pines so the stems taper inward near the base. This makes them look especially precarious, like a telephone pole balanced on top of a pool cue. The trees are so brittle and rotten we hardly have to use our saws. We get some momentum going with a couple of pushes and the trees just fall over, no chainsaw required. I would do this for fun in my spare time.

FALLING CONTINUES THE next day in a stand of sickly trees on thin soil. We walk into the section; there are trees leaning over and tangled up in each other everywhere. When we get into the middle of our falling area, the wind picks up and a squall moves through.

Trees start coming down like I've never seen before. They're limbless save for their very tops, where wind catches the foliage and pushes the tree over. The strength of the steady gusts is a chokehold putting them to sleep. Sometimes one leaning tree will nudge up against another, causing both to shudder as it comes to rest. Sometimes they fall in stages, taking minutes to hit the ground, minutes brought to life by the roots snapping and popping like metal in the hot sun. Some fall so slowly you could lie underneath them and they wouldn't even kill you, they'd just press you into the moss.

At first I'm calm as the trees fall. But suddenly a mess of wood, bent horizontal and cribbed into the trees above us, comes down in a rush of a hundred machine-gun snaps. Trees caught in the nest flail around before hitting the ground. Our eyes dart everywhere, trying to keep track of every moment. Trees break free and swing themselves like catapults. Splintered chunks of wood slash through the air like propellers. Tabes and I look in opposite directions, standing guard for each other. We don't dare move, as that would take our complete focus from the storm of debris.

As the mess is coming down, the rumblings in the ground let loose a second wave of tree-falling fury. It starts close then cascades away from us like the breaking of an ice jam. We watch the destruction. Nothing hits us.

As soon as it's over I turn to Tabes. "Fuck this, let's get out of here."

We go back to the guard and watch as dozens more trees fall over.

We're called down to where the rest of the crew is working. There was an escape across the guard last night. It rained in the morning, but strong winds are pushing the fire and suddenly we're dealing with fire weather more common on July 7 than September 7.

Tabes and I go into the escape to fall. The work is endless and nearly as dangerous as it was in the last spot. We're working among bigger spruce trees down here and the winds are causing them to fall on their own.

We look around the spot—hot ash everywhere and a wall of smoke. We're not sure where to start, since everything is about to fall anyway. After some looking and talking we dig in.

Adrenalin percolates in my body and I operate my saw in a state of mild panic.

An hour or so after Tabes and I part ways to work in different areas, I hear a tree crash down. Over the din of the fire I hear somebody roar, "FUCK."

I freeze, terrified that Tabes has been hurt. There's another shout, "Damn it!" with a strange tone. He's obviously not dead; at worst he's injured. I'll call on the radio but decide to give him a few seconds to cool down.

"Ranger Charlie Two, it's Ranger Charlie."

Seconds go by. I plan the moves to make if Tabes is hurt.

"HEY!" The shout comes from behind me. Tabes is walking toward me; the last bits of fear and shock on his face are being replaced by anger and relief.

"I just about died."

"Shit, man, what happened?"

"Tree came down, didn't hear it or see it or nothin'. Hit my bar as I was cutting another tree."

"Jesus."

A tree hitting your chainsaw is as close as it gets.

We walk over to the spot where he was falling. With some easy detective work, we see where a tree burning halfway up the stem broke off and cruised through the air in silence, hitting Tabes's saw.

Falling trees is one of the most dangerous jobs in North America. It feels true today. This is also the first day this summer I've felt real fear exhale hot on my neck. It's the kind of fear that tires you out and I'm destroyed by the time we pack up for camp. First the jumble of falling trees in the morning, then the close call with Tabes in the afternoon. And for what? Timber value?

When we get to the fire line the following morning, the weather is low overcast. Back at camp it's completely fogged in and the medevac helicopter can't fly. If we can't be flown out of the bush in case of injury, we can't work.

Some other time I'd say that's lame, but after yesterday's close call it seems sensible.

We wait at the trucks. People read or sharpen Pulaskis or do push-ups. Nobody is antsy. We're fatigued, but the fatigue has been around long enough to feel normal.

There used to be ups and downs when we had to wait like this. On our first fire, somebody cranked "Bohemian Rhapsody" and people screamed the lyrics. It's embarrassing thinking back on those days earlier this summer, like the first week of college residence.

This morning shows a different group, as flat and static as the grey clouds above us.

Tabes sits in the back of the truck staring out the window, fine to go to work, fine to sit there. His expression wouldn't change if apes riding grizzlies ran out of the bush and attacked.

Somehow, it feels like we're farther away from humanity right now. Physically that's not true. We're the same distance from civilization as on earlier deployments: five hours' drive from the nearest 7-Eleven, which equals civilization in northern BC.

But you can get farther than physical distance. Your mind can wander past whatever remote, burnt-out forest you're tending. Parts of your brain dedicated to being a functioning member of society calcify and die off. Other parts come alive, parts that can hear changes in a bird's song or note the tiniest shift in timber type.

When it's time to actually go to work everyone in the truck jumps out of their seats in unison. All the doors slam at the same time. Packs are mounted, gear is loaded and we're walking down the hill in a silent, telepathic pack.

I remember going to my dad's logging operation last fall and seeing the same thing. His crew took a small boat to work. When they got to the job site, they walked down the dock and jumped into their trucks. Dispersed to their proper places. None eager. All with purpose. Few of those guys love their jobs, but that's what they did. Repetition as necessary as the sun.

Maybe that's why I was so flustered a few nights earlier, talking to Dan, at the thought of never fire-fighting again. Life is a rat's nest of questions, but work soothes them.

LATER THAT AFTERNOON, a call comes over the radio.

"All stations, all stations" is how it starts, not something you want to hear on the radio. It means there's been an accident and that all operations, including radio chatter, have to be shut down.

Somebody on another crew has been hit by a tree. I'm furious. I swear under my breath. I do this alone and keep calm as we gather at the guard to see how we can help. The other crew is the Monashee unit crew from the southeastern part of BC. They're working just up the road from us. They'll need help getting the injured worker out of the bush.

Warren's squad is working closest to Monashee and the site of the accident, so they leave right away to help.

When the rest of the group is gathered, Dan starts giving orders. First-aiders go right away; the rest of us will collect our gear and start walking up.

Walking up the guard I think about the situation. The Monashee unit crew leader, a woman in her early thirties, sounded calm on the radio. But I don't know her well enough to gauge her tone and speculate about the accident. I assume the worst. The trees are small where they're working, and that makes me hopeful. But this can be deceiving. "We're the softest thing in the bush," Dan likes to say.

When we're all in the truck we take off up the bumpy road. Silence as Tabes drives. We get to the spot where everyone is parked and it's a rush of activity done without words. The silence is broken only with instructions. As soon as we have enough words to understand tasks, we do them. Dan is directing.

"Lauren, water the shit out of this helipad. Tabes, Willy, check for danger trees at this pad and make sure there's nothing funky."

Warren's group is already in the bush. Tom is helping cut a trail to the injured worker so they have a path wide enough to pack the stretcher out. Other squad members are helping get the faller onto a stretcher. Warren, our most experienced first-aider, is overseeing it all.

A group of us waits at the truck for the stretcher to come out of the bush. Everyone looks at the ground or into the distance. The group comes out of the bush with the stretcher, three on each side and a steady low babble of communication between them warning of coming twists or dips in the trail. Warren walks behind, radio in hand. They look small as they wind their way through the trees.

I'm relieved to see the guy on the stretcher awake and, other than being on a stretcher, looking fine.

Warren tells me the worker broke a collarbone and a kneecap and was showing signs of shock but was okay overall.

"It went about as fluid as you would expect," Warren

says of the response to the accident. "You know, those things take over"—"those things" being a lot of first-aid drills done when we're on base.

We load the Monashee crew member into the box of a pickup truck. He's taken up the hill to a waiting helicopter, where he'll be flown to the hospital in Prince George.

When we get back in the evening Tabes talks with Dan about how we weren't covered by a medevac helicopter for a few hours yesterday. We hadn't realized this until later in the day, after we'd started working. Tabes is worried that Dan was so eager to get to work that he decided to ignore it. Dan tells him that wasn't the case, that it was a simple miscommunication.

Now isn't the right time to talk about it, but it seems like the only time to talk about it. Their conversation is strained and I stand silently while they talk. Dan asks me to weigh in and I take my time. I'm angry too, wiped out by today's events and yesterday's, and this whole summer in the bush.

I want to explode at Dan. To me he seems dismissive of what's happened in the past two days. Somebody on another crew was hit by a tree. Tabes was inches from the same accident. But Dan doesn't appear too concerned. We should wait several months to talk. Yet here we are, sorting it out still in the heat of the day's events. I speak as calmly as possible, but my voice still shakes. We get through it; we shake hands and go our separate ways. Still, nothing seems clear.

I learn later that after they carried the faller out on the stretcher, the safety officer for the fire walked in to the spot where the accident happened. He took pictures of the tree that had caused the injury, but it was hard to understand what had happened because three more trees had already fallen on top of the first one.

ON RETURNING TO our tents we find a big membranous tube funnelling sweet, hot air into our cold living space. It's a welcome change from nights that keep getting colder. Everyone entering gives a throaty "ohhhhh" when they realize what's happened.

The heat has come on just in time. It's –10°C in camp the next morning. Negative ten and we're still firefighting. All the pipes are frozen; even the septic tank has given up wafting its rancid fumes.

Before we can go to work we have to attend a meeting with the entire camp about yesterday's incident. Everyone stays in their heated tents or trucks until the last minute. Then at seven a.m. all four unit crews descend on the meeting area. The message comes from the fire's Incident Commander, a feisty man with grey whiskers named Rob. In the cold morning he wears a blue trench coat with hi-viz stripes. Rob makes it quick.

"Someone got hurt. Be careful. Now go back to your trucks, it's freezing out here."

Before we can work, an excavator has to fix a mud pit in the road. I'm still coming down off yesterday's drama— the tough conversation with Dan and the discomfort of seeing one of our own get packed out of the bush. I'm questioning why we're out here. Somebody almost dies and we just keep going. Firefighting often feels like an excuse to goof around with your buddies in the bush, not a sometimes dangerous vocation.

The sun is out and hanging with the crew is

therapeutic. But eventually the road is made passable and we have to go to work.

When we get to the fire we see that, like the pipes in camp, our hose lines are frozen. We disconnect the hose and step on the lines; Freezie-sized tubes of brownish ice ooze out of the ends. The sun is still well below the trees, so we get to work grubbing. It's useless busywork—there's too much fire for us to be digging around, but we have to wait for water in liquid form.

Despite the sun and the good group of people, we're all frustrated at the situation. We dig through boulders trying to get at heat. If we find anything warm, we mix it with soil to put it out. The task is never-ending. You pull out a rock and the smouldering embers fall into the crack of another rock. As we work we talk about what we want to do with our lives, our ideal careers. Blaine is especially hung up on his future today. His body twists in discomfort as we ask him questions about the coming year, whether he'll go to school or work or travel. He's indecisive. Blaine has been with the crew for six seasons now, but he started young and is still heavily influenced by the advice and ribbing of his crew mates. At this late stage in the season, with decisions looming, getting into his head is easy.

After an uninspired half-hour of work I reach for my lunch kit. I open it and squat over it, pawing through the assortment of plastic-wrapped carbs. I eat like a scavenger consuming a carcass. Finishing my peanut butter and jam on white bread, I look up the road and see Blaine squatting down and sifting through warm, fire-stained boulders, just as much an ape as I am. He's on the edge of the guard, silhouetted in the morning sun with the smoking forest behind him. He stops sifting and gets to his feet; he looks ahead and his shoulders drop as he lets out a deep sigh.

THE DAYS, ONCE marked by changing tasks and fifteen-metre flames, are now coagulating and piling into each other. We're camp people—waking up at the same time, eating with the same people, going to bed at the same time. Routines get so entrenched they alter your view of the world. We're becoming the type of unreasonable bush workers who would throw a fit if the cooks were to remove bacon from the breakfast menu one morning.

But then we're thrown a life ring. We're moved to a new spot on the fire the next day.

Our first task at the new spot is to sit and wait for instructions. I have some cigarettes from an ex-Ranger friend who bought them on a trip to Brazil. This seems like a good opportunity for a smoke.

I jump out of the truck and a few others join me. There's no filter between the tobacco and our lungs, which could be scraped and used to fill potholes by now. As we stand there smoking we all get a little dizzy, a reprieve from whatever it is we're doing out here.

Dan comes back from a flight with instructions. Six of us are sent to put out hot spots identified in a scan. Scanning is done with an expensive, gun-like piece of equipment that can detect heat. In the later stages of a fire, a helicopter will go out early in the morning, scan the edge of the fire and mark the hot spots using GPS coordinates.

All afternoon we grub at these hot spots. Many could

use water instead of just tools. We dig at one spot for almost two hours. The ground is rocky and our Pulaskis are dull as butter knives from weeks of chipping away at rocks.

Once it looks like we've killed all the smoke, we sit and wait for our piles of hot ash mixed with dirt to start breathing again. Nobody talks during the rests; our low energy levels are synced. We dig, sit, doze off, dig again and walk.

I'M TOLD TO choose somebody to work with the next day and I go with Brad, a safe choice. This is a guy who's known to say things like, "Sometimes you just have to work for the sake of working."

We go back to all the spots we hit yesterday and there's still heat in every one of them. New spots have been identified by a scan as well. All the spots are in berms—piles of dirt mixed with sticks and other organics. Berms are usually left over from logging or road building. They look like the piles you might see at the edge of a freshly cleared lot. On the Good Firefighting Jobs scale, digging in berms ranks near the bottom.

The extremely high drought codes mean this fire is buried deep in the dirt. These berms are nearly impossible to extinguish. Instead of a single, easy-to-see blue wisp, smoke comes out of them like sweat through pores.

There are several methods for finding out if a berm is hot. There's looking (not great), touching (sometimes burns) and sifting (best). Sifting means a couple of whacks at the berm with your Pulaski, a cure-all for most things in firefighting.

We pull up to one of the berms. It's in the wide-open

space of a safety zone. After we start digging I go back to the truck and flip the satellite radio to Ozzy's Boneyard, a grinding hard-rock channel. We listen to Danzig and the Cult and Mötley Crüe as we dig. The dried-out piles of hot dirt we destroy match the texture of the music. Ozzy would be proud.

Another spot we're looking for is in a creek draw. Next to the creek is the biggest Interior spruce tree we've ever seen. Brad and I see if we can touch arms around it. Our hands connect, barely. It feels strange; this move is for yuppies walking the West Coast Trail, not men who listen to Ozzy's Boneyard.

ON OUR LAST line day, Brad and I go on a search for two more spots, leftovers from yesterday's mission. The rest of the crew is hosing high on an old logging road at the base of a steep ridge. The spots Brad and I are looking for are on the other side of the ridge.

There are dozens of roads and fire guards in this area. Any of them could lead us to the other side of the ridge. We try several but the best we can do is get within a couple of kilometres of where we need to be. The GPS ticks us toward our target. Distance to spot: 2.02 kilometres, 2.01 kilometres, 2.00 kilometres, 2.01 kilometres.

"Stop, we're getting farther away," I say to Brad.

We park and shoulder our bags. The temperature is in the high twenties. We walk toward our next spot. Probably a berm.

We're directly on the other side of the ridge from the rest of the crew. We cross a steep-banked stream and connect to another logging road. It's all fresh logging up here, two years old at most. Big swaths of the hillside have been removed, but as the ground gets steep the logging blocks end. Steep hills thin the already-thin profit margins of logging the northern BC Interior.

The recent logging, the way the roads are still smooth, bridges brand new, adds to the seclusion. It's jarring the way this place was so cleanly devoured, first by logging, then by fire, then left alone.

We walk the road until our spot fire starts getting farther away again on the GPS. But it doesn't matter. The edge of the Chelaslie River fire is now a point in time. It won't exist until the season changes, until this northern latitude is catching less sun and more rain.

We start walking back to the truck. The conversation turns to women. It's all carnal. We aren't discussing which women from our past were cool or sweet or romantic. It's all about who was the hottest. We do it because we're lonely and bored, and because it's cathartic.

Just before we get to the trucks, a helicopter carrying Dan, Kelly and Lauren appears out of the gathering smoke. Via radio, we explain the situation to Dan. He tells us to wait.

Ten minutes later, the three of them arrive and the chase is back on. We hoof it up the road. The only sounds are boots and breathing. We get to the same spots we saw from a distance, but now there are five of us, and there's a helicopter coming to bucket these spots.

Our two spots are a berm and a patch of fire on the green side of the road with lots of open flame.

The helicopter comes into focus out of the white-grey smoke of the sky, which is now blocking our view across the valley. It buckets the berm and we go up the road to the spot fire. It's burning through slash at a good clip, and it still looks like there's no way we'll put it out. But we start in with our tools and it starts disappearing.

We continue chopping and scraping, working in true last-day form—overly tenacious, not very efficient. I dig in fits, going at a furious pace for a couple of minutes, then backing off and taking a breather.

After an hour I eat half of a marshmallow square. After another hour I eat a PB and J on white, using extra packets of peanut butter and jam to maximize edibility.

It's time to leave as I'm finishing my sandwich. My exit is reluctant. I know there's still heat. But I leave. This tour is over. Fine.

The smoke is sitting down fast now. By the time we're at the end-of-day meeting, we're shrouded in the thick dark fog that's defined these last two months.

When we get back to camp, I get out of the truck with the agility of a block of concrete. My back is sore, as is my neck. My hamstrings are tight. I limp to the tent, sit down on my cot and untie my boots. I grab the worn leather and pry them from my feet. My exposed socks smell like compost.

The rest of the crew is ahead of me, gone to get dinner. I'm alone in the tent. The sun is setting behind the smoke outside. Inside it's almost dark. The cream-coloured canvas of the tent filters the already soft glow of the evening light.

I pull off my red uniform shirt. The shirt is sticky with sweat and halfway up my body it gets stuck, bunching up at the armpits. My face is held tight against the damp fabric of my cotton undershirt. I try to fight it off at first. Then I stop.

It's hot in my shirt; it smells like sweat and dust. Like

work. It smells like my dad, and his dad, and his dad before him. It smells like my childhood. It smells like the laundry room in my grandmother's basement. A smell of metal and dirt and gas.

I grab at the coarse fabric of my shirt until it starts moving up my back. I pull the shirt off and throw it on the floor next to my cot. Then I put on a sweater and go to the dinner line.

Epilogue

When we return to the base we learn that crews are still being sent to work on the Chelaslie River fire. We also learn we will not be one of those crews. Dan takes on the bulk of the frustrations with these developments.

Our frustration stems from the fact that these crews are from the farthest southern reaches of the province. We feel (perhaps irrationally) that if the fire is in our backyard, we should be the ones working it to the end. When we're down south, we're usually the first to be sent home as things start slowing down.

I'm mostly detached from the situation, too burnt out to feel indignant. I'm away from the crew all day writing evaluations for departing crew members. While I write, they're out detailing the trucks. I walk by a couple of times but don't say much, as they're distracted by loud music and thorough polishing. Just as well because every time I open my mouth it's some lament about it all being over too soon.

In the crew trailer I go over the evals with Dan. I'm not interested in being too critical of anyone and even Dan, usually a hard marker, talks himself out of most of his criticism.

"We don't want so-and-so to leave with a bad taste in their mouth," he says.

One night after work I go to the movies with Blaine and Addison. When we come out after the show it's dark and raining. Pounding rain, fall rain, season-ending rain. We linger under the awning in front of the theatre for a while before getting in our trucks and driving home. This is the first time rain has felt good all year. It's killing the fires and the season, and I'm glad of it.

As I'm driving home I imagine I'm on a street in Halifax, where rain will just be rain and it won't affect my paycheque or my prospects for adventure. Sometimes Sue and I go out for a walk on these sorts of miserable nights. She always feigns dread when I suggest it.

I'd love to see the Chelaslie fire right now—the heat mixing with moisture for hundreds of kilometres creating a steam bath for the gods, bleeding the heat dry, finishing the battle for good.

The next night we go out for an end-of-the-year gathering. Dan, Warren, Rob and I have to stay on longer because our contracts aren't up until the end of October, but for everyone else this is it. Their season is over. We go to the Smithers pool after work, then we go to the pub. Everyone is getting pretty drunk. Everyone save for Dan and me.

We continue bar-hopping and the group gets split up. One person gets kicked out of a bar, while another gets lost in transition between two bars, wandering down a dark side street shrouded in spruce trees.

Dan is going to drive me home, but before we leave we attempt to round up a few strays to make sure they're all right.

We're at the intersection of Main Street and Highway 16 and Dan pulls off to the side. The group should be around here somewhere. There's no wind and the night air is thick with the smell of yesterday's rain.

It's silent in the truck, silent but comfortable. Maybe this is Dan's mythic cowboys-on-the-ridge moment. Or maybe we're just tired. It's been our longest summer together and this might be our last mission—to check on the crew once more before it's no longer his problem and by extension no longer my problem.

Then we see them. They're jogging down the highway. Blaine is in the lead; he's pushing a stolen shopping cart and Addison is stuffed in the cart, his limbs draped over the sides. As they approach the intersection we hear the cart's wheels humming along the pavement and Addison yelling "Ahhhhh" as they run through the heart of downtown Smithers. Behind them is the rest of the crew, jogging and shouting and cradling McDonald's cups and half-eaten boxes of french fries.

The scene recedes out the side window. Everyone is accounted for.

I SUPPOSE I'D known since May that this would be my last season on the Rangers. I remember sitting on the tailgate of one of the trucks in early October. I had been doing some cleanup work on the China Nose fire. I was given a map and paired with an excavator operator. We drove to various spots on the fire, doing things like pulling debris back across cat guards and making sure creeks and streams hadn't been disturbed. Not much was demanded of me; I mostly just walked around and watched the excavator work.

On one stretch of road there were some berms still smouldering. I whacked away at them with my Pulaski for a while, just to say I'd worked some smokes in October. Later that afternoon we ran out of work. Our last bit of rehab was pulling a few muddy trees out of a ditch where a cat guard had punched in off a logging road. We were at a high elevation and the cold hurt my ears. It rained a few fat drops every once in a while. I said goodbye to the operator and he drove down off the mountain. I went back to the truck to pack up for the day. There was nothing left of the China Nose fire. Every helicopter gone, every pump, every chainsaw, every length of hose and every red shirt. All that remained was me and a green Ministry truck and a million blackened trees. I threw my hard hat and Pulaski in the box and sat there on the tailgate for a long time. Then I drove home.

Acknowledgements

THIS BOOK LIKELY wouldn't exist if Brianna Cerkiewicz, Howard White and Anna Comfort O'Keeffe at Harbour Publishing hadn't seen some potential.

Thanks to everyone at King's College in Halifax for all the help and for sharpening up the writing—Don Sedgwick, Stephen Kimber, Lori A. May, David Hayes and Ken McGoogan.

Thanks to Dan Dykens for being a friend and leader and to Sue Pearce for answering many questions and for the title. Thanks also to Andrew Pearce, Brendan Hutchinson, Cathy L'Orsa and Levi Froese.

Thanks to my parents, Kelly and Lauren Williams, and my sister Sarah, all great people. And Sue L'Orsa, you can't be thanked enough.